Optical Spectroscopy

Fundamentals and Advanced Applications

Other Related Titles from World Scientific

The Concise Handbook of Analytical Spectroscopy: Theory, Applications, and Reference Materials
(In 5 Volumes)
Volume 1: Ultraviolet Spectroscopy
Volume 2: Visible Spectroscopy
Volume 3: Near Infrared Spectroscopy
Volume 4: Infrared Spectroscopy
Volume 5: Raman Spectroscopy
by Jerry Workman
ISBN: 978-981-4508-05-6 (Set)
ISBN: 978-981-4508-08-7 (Vol. 1)
ISBN: 978-981-4508-09-4 (Vol. 2)
ISBN: 978-981-4508-10-0 (Vol. 3)
ISBN: 978-981-4508-11-7 (Vol. 4)
ISBN: 978-981-4508-12-4 (Vol. 5)

Optical Properties and Spectroscopy of Nanomaterials
by Jin Zhong Zhang
ISBN: 978-981-283-664-9
ISBN: 978-981-283-665-6 (pbk)

Photosynthesis: Solar Energy for Life
by Dmitry Shevela, Lars Olof Björn and Govindjee
ISBN: 978-981-3223-10-3

Optical Spectroscopy

Fundamentals and Advanced Applications

Emil Roduner
University of Stuttgart, Germany,
and University of Pretoria, South Africa

Tjaart Krüger
University of Pretoria, South Africa

Patricia Forbes
University of Pretoria, South Africa

Katharina Kress
University of Stuttgart, Germany

World Scientific

NEW JERSEY · LONDON · SINGAPORE · BEIJING · SHANGHAI · HONG KONG · TAIPEI · CHENNAI · TOKYO

Published by

World Scientific Publishing Europe Ltd.

57 Shelton Street, Covent Garden, London WC2H 9HE

Head office: 5 Toh Tuck Link, Singapore 596224

USA office: 27 Warren Street, Suite 401-402, Hackensack, NJ 07601

Library of Congress Cataloging-in-Publication Data
Names: Roduner, E. (Emil), 1947–   author. | Forbes, Patricia B. C., author. |
     Krüger, Tjaart, 1983–   author. | Kress, Katharina, author.
Title: Optical spectroscopy : fundamentals and advanced applications / by Emil Roduner
     (University of Stuttgart, Germany & University of Pretoria, South Africa), Patricia Forbes
     (University of Pretoria, South Africa), Tjaart Krüger (University of Pretoria, South Africa) and
     Katharina Kress (University of Stuttgart, Germany).
Description: New Jersey : World Scientific, 2018. | Includes bibliographical references.
Identifiers: LCCN 2018039248 | ISBN 9781786346100 (hc : alk. paper)
Subjects: LCSH: Optical spectroscopy. | Spectrum analysis.
Classification: LCC QC454.O66 O684 2018 | DDC 543/.5--dc23
LC record available at https://lccn.loc.gov/2018039248

British Library Cataloguing-in-Publication Data
A catalogue record for this book is available from the British Library.

Cover images show fluorescence microscopy images of a brainbow of mouse neurons (reprinted from Smith, 2007, (https://en.wikipedia.org/wiki/Brainbow, downloaded 05 June 2018) and bovine pulmonary artery endothelial cells (https://en.wikipedia.org/wiki/Fluorescence_microscope, downloaded 05 June 2018).

First published 2019 (hardcover)
Reprinted 2024 (in paperback edition)
ISBN 9781800616349 (pbk)

For any available supplementary material, please visit
https://www.worldscientific.com/worldscibooks/10.1142/Q0182#t=suppl

Desk Editors: Herbert Moses/Jennifer Brough/Shi Ying Koe

Typeset by Stallion Press
Email: enquiries@stallionpress.com

Preface

Textbooks tend to lag notoriously behind what is needed in the real world of science and its applications. This is perhaps understandable since in science there is often an established academic syllabus that is largely similar and unofficially standardised around the world. Furthermore, new directions first have to be established and their fundamentals systematically digested for presentation in classrooms.

The fields of central interest in science shift periodically as new subjects come up or new technological methods become available. Over the past several decades there has been significant progress in several fields of spectroscopy from long wavelength NMR spectroscopy for structural studies of solids and of macromolecules in solution to optical spectroscopy in the ultraviolet and visible part of the spectrum to short wave X-ray spectroscopic methods.

The present work is written as an advanced textbook on optical spectroscopy in the range of ultraviolet and visible radiation with the aim to close the gap between the existing standard textbooks in physical chemistry and the present demand in research and its applications in biology and materials science, including light-harvesting systems in energy applications for solar cells, solar water splitting and natural or artificial photosynthesis. It is intended as a basis for graduate courses in chemistry, biology and materials science and for scientists who want to start using optical spectroscopy for their research. Rather than focusing on a small slice of the subject in full depth it intends to provide an overview of the subject from atoms to molecules, semiconductor quantum dots to bio-matter, in the hope that cross-fertilisation of well-documented phenomena in one field to other fields will aid understanding. More in-depth details on topics are made accessible via recommended further reading.

The work was triggered first by my former student Roland Heugel in Stuttgart whom I thank for encouraging me repeatedly to write a textbook on spectroscopy. Secondly, the PhD thesis of Katharina Kress for which I served as an unofficial co-supervisor turned out to become a systematic, textbook-like work that illustrates perfectly the optical properties of organic chromophores. I am grateful that Katharina accepted to be a co-author in charge of Chapter 4. I am just as grateful to my colleagues Patricia Forbes from the Chemistry Department and Tjaart Krüger from the Physics Department of the University of Pretoria for contributing their expertise with colloidal quantum dots (Chapter 5, Patricia Forbes), instrumentation and data analysis (Chapter 3, Tjaart Krüger) and applications to biological problems, particularly photosynthesis. Both these co-authors contributed to the applications in Chapter 7. Last but not least, I thank my dear wife Hanny for her patience when I was engaged with my manuscript.

Emil Roduner
Stuttgart/Pretoria
April 2018

About the Authors

Emil Roduner is a former Chair in Physical Chemistry at the University of Stuttgart, at present part-time at the University of Pretoria. His broad interests include muonium chemistry, elementary steps of catalytic reactions, kinetics and dynamics of free radicals, magnetism of metal clusters in zeolites, degradation and proton conductivity distribution of fuel cell membranes.

Tjaart Krüger is an Associate Professor in Biophysics at the University of Pretoria. His main interest is to resolve the molecular details of energy transfer and regulation in the light-harvesting complexes of photosynthetic organisms and to control these processes using shaped light and nanoparticles. He uses various optical spectroscopy techniques and designs and builds specialised spectroscopy instruments.

Patricia Forbes is an Associate Professor in Analytical Chemistry at the University of Pretoria. Her research focuses on the development of novel sampling and analytical methods for environmental pollutants, including denuder-based sampling techniques, biomonitors and quantum dot-based fluorescence sensors.

Katharina Kress is a former PhD Student from the University of Stuttgart. Her thesis focused on a systematic study of organic dyes. She wanted to understand the correlation between the structure of a dye and the observed spectroscopic phenomena. Further calculations and possible applications of these molecules complemented her studies.

Contents

Chapter 1

Introduction

The advent of optical spectroscopy is intimately related to the development of quantum mechanics. This is a consequence of the absorption lines that were first observed in the visible part of the solar spectrum by Wollaston (1802) and rediscovered independently by Fraunhofer (1814). It was recognised about 45 years later by Kirchhoff and Bunsen that the lines coincide with characteristic emission lines of heated elements. At the time, they were simply accepted as empirical facts.

Of greater impact were the emission lines observed upon a discharge in hydrogen gas. It was understood that line spectra were in conflict with classical physics that assumes continuous energy distributions and does not account for energy quantisation. The Swiss mathematics teacher Johann Jakob Balmer (1885) found a simple formula that can reproduce the wavelengths of the lines empirically. Understanding that the lines represent transitions between discrete energy states of atomic hydrogen led to the development of the Bohr model (1913), shortly after Rutherford had found that the atomic mass is mostly concentrated in a nucleus, whereas the electrons have little mass but take up all the space (1911). The Bohr model reproduces accurately the Balmer formula, but it describes a flat (2-dimensional) atom in which the ground state has an orbital angular momentum. The proper 3-dimensional model was found as a solution of the time-independent Schrödinger equation (1925) and extended by Dirac to include relativistic effects (1928).

It was soon recognised that molecules also display line or band spectra, reflecting the fact that energy quantisation is ruling also the electronic states of molecules. The first UV absorption and fluorescence spectra were recorded 1922 with benzene by Victor Henri at the University of Zürich, using an electrical discharge in water as the UV source. Zürich was one of the key places where quantum mechanics was developed, and Henri was in close contact with Hermann Weyl, Erwin Schrödinger and Peter Debye. He applied quantum theory to the high-resolution line spectra of gas phase benzene, obtaining the moments of inertia from the rotational fine structure. In 1928, he interpreted the line broadening in spectra of gaseous formaldehyde to the decay of the molecule and thus discovered predissociation. He wrote an early textbook that takes account of the quantum-mechanical description only 10 years after its discovery.[1]

High-resolution spectroscopy revealing detailed understanding of the structure of small gas phase molecules continued to be at the centre of interest of optical spectroscopy for several decades. Key contributions came from Gerhard Herzberg who received the Nobel Prize in Chemistry 1971 "for his contribution to the knowledge of electronic structure and geometry of molecules, particularly free radicals". His classical books which are of fundamental use also to chemistry in interstellar space include the fundamentals of spectroscopy in the UV/Vis, Raman and infrared and tabulated properties.[2-7]

Still today, high-resolution spectroscopy dominates the teaching syllabus in physical chemistry, which has led to various further fundamental textbooks.[8,9] In recent decades, the developments have taken a different direction with the focus shifting away from understanding small gas phase molecules towards applications in materials and in biological systems. This is a consequence of the increased interest in these topics but also in technological developments. Much progress has been achieved in technology by the development of lasers as powerful monochromatic light sources. They can be combined and pulsed, permitting the use of higher order effects and ultra-high time resolutions, down to the femtosecond range.

The unprecedented sensitivity of photodetectors that can count single photons combined with the spatial resolution of confocal microscopes allow the observation of single molecule fluorescence in highly diluted samples, even in biological cells. This permits the study of properties of individual molecules in their often inhomogeneous environments. It contrasts with measurements of ensembles, which always provide averaged values and in the case of overlapping bands of mixtures does not distinguish between different chromophores. Fluorescence lifetime and polarisation can be measured, translational and rotational diffusion observed and intermolecular distances inferred. Luminescence of molecular or colloidal quantum dot markers may be intermittent (on–off or blinking behaviour). A key book on the Principles of Fluorescence Spectroscopy was written by J. R. Lakowicz.[10]

Absorption, luminescence and excited state energy transfer properties have become of crucial importance on a large scale in materials related to light harvesting in organic and inorganic (perovskite or quantum dot) third generation solar cells, for solar water splitting and in light emitting diodes, TV screens and many other applications. One of the key developments is based on the early work of Theodor Förster who discovered a new energy transfer process of electronically excited states.[11] Its efficiency has a pronounced distance dependence that is used today as a ruler for the characterisation of the structure and dynamics of DNA, proteins and other biomolecules which are site-specifically labelled with fluorescing markers.

The present book takes into account these new developments and covers the basics of absorption, emission and energy transfer of molecular systems in the condensed phase as well as the corresponding behaviour of metal nanoparticles and semiconductor quantum dots. Following the rule ascribed to Stephen Hawking that each mathematical equation in a text reduces the number of its readers by 50%, we keep the mathematical formalism to a minimum, apart from the chapter on data analysis where equations are unavoidable. Instead, we attempt to address graphical recognition by including a large number of didactic colour figures. The book explains the fundamental phenomena based on an instructive set of spectra and their dependences on various parameters observed with rigid merocyanine dyes and extends these principles to semiconductor quantum dots. It explains the experimental setup and the data analysis that is often tedious because of the interference of unwanted signals from other sources in the sample and the spectrometer. It then treats energy transfer processes of excited states and finally focusses on a set of key applications, notably in biological systems, including photosynthesis, solar cells and solar water splitting systems. These applications all build on a sound understanding of the optical properties and the processes following optical excitation.

References

1. V. Henri, *Physique moléculaire: Matière et énergie*, Hermann et Cie. Paris, 1933.
2. G. Herzberg, *Atomic Spectra and Atomic Structure*, Dover Books, New York, 2010, ISBN: 0-486-60115-3.
3. G. Herzberg, *The Spectra and Structures of Simple Free Radicals: An Introduction to Molecular Spectroscopy*, Dover Books, New York, 1971, ISBN: 0-486-65821-X.
4. G. Herzberg, *Molecular Spectra and Molecular Structure: I. Spectra of Diatomic Molecules*, Krieger Publishing Company, Malabar, Florida, USA, 1989, ISBN: 0-89464-268-5.
5. G. Herzberg, *Molecular Spectra and Molecular Structure: II. Infrared and Raman Spectra of Polyatomic Molecules*, Krieger Publishing Company, Malabar, Florida, USA, 1989, ISBN: 0-89464-269-3.
6. G. Herzberg, *Molecular Spectra and Molecular Structure: III. Electronic Spectra and Electronic Structure of Polyatomic Molecules*, Krieger Publishing Company, Malabar, Florida, USA, 1989, ISBN: 0-89464-270-7.

7. K. P. Huber and G. Herzberg, *Molecular Spectra and Molecular Structure: IV. Constants of Diatomic Molecules*, Van Nostrand Reinhold Company, New York, 1979, ISBN: 0-442-23394-9.
8. J. M. Hollas, *Modern Spectroscopy*, 4th Ed., John Wiley & Sons, Chichester, UK, 2005, ISBN-13: 978-0470844168.
9. C. N. Banwell, E. McCash, *Fundamentals of Molecular Spectroscopy*, 4th Ed., Mc Graw Hill, New York, 1994, ISBN-13: 9780077079765.
10. J. R. Lakowicz, *Principles of Fluorescence Spectroscopy*, 3rd Ed., Springer, Singapore, 2006, ISBN-13: 978-0387312781.
11. T. Förster, *Ann. Physik.*, 1948, 437, 55–75.

Chapter 2

Fundamentals

2.1 The Nature of Light

Light is an oscillating electromagnetic field that has wave nature with wavelengths λ extending from 370 to 730 nm (nanometre, 1 nm = 10^{-9} m).[1] This λ range is visible to the human eye and is but a very small portion of the electromagnetic spectrum, which extends over many orders of magnitude, from below 1 picometre (1 pm = 10^{-12} m) for gamma-rays to radio waves with λ up to 100 km and more. Optical spectroscopy, which is the subject of this book, describes the interaction between light and matter, where a broad λ region for "light" is often considered that includes visible light and the portion of ultraviolet (UV) light that extends from 400 nm down to about 180 nm. It is used to probe the structure and properties of matter.

Electromagnetic radiation can propagate through vacuum and through any kind of matter that is transparent to it. In vacuum, it travels at $c = 2.9979 \times 10^8$ m s^{-1}, known as the "speed of light", which is about the length of a ruler (30 cm) in 1 ns. In matter, its speed is slowed down by the index of refraction of the material. The index of refraction of vacuum is $n = 1$, and any other material has $n > 1$.

Unlike sound waves, electromagnetic waves have a dual nature and in addition to their wave character they exhibit also

7

the character of particles with rest mass zero. These particles are called photons. The intensity of light varies depending on the number of photons. Each of these carries an energy $E = h\nu$, where $h = 6.6261 \times 10^{-34}$ m^2 kg s^{-1} is Planck's constant. The important relation to remember that connects energy E, frequency ν and wavelength λ is

$$\Delta E = h\nu = \frac{hc}{\lambda} = \frac{1239.8 \text{ eV} \cdot \text{nm}}{\lambda \, [\text{nm}]}. \tag{2.1}$$

Thus, energy scales proportionally with ν but inversely with λ. Bright light consists of many photons, and if the light is monochromatic all of them have the same frequency and wavelength, i.e. the same colour, hence the name monochromatic. Photons can be counted, and the unit of 1 mol of photons is 1 Einstein. Due to its particle nature, the energy of light is quantised, meaning that it comes in portions, similar to sugar when it is delivered as sugar cubes. In contrast, sound waves of a certain wavelength can adopt a continuous range of energies and weaken gradually.

Photons carry no rest mass, no charge and they are stable particles. However, even though they have no mass they carry a linear momentum p, given by the de Broglie relation, $p = h/\lambda$. They also carry angular momentum, normally characterised by the orbital angular momentum quantum number ℓ and magnitude $\pm\hbar$ for circularly polarised light, where $\hbar = h/2\pi$. This is important since angular momentum is conserved in spectroscopic transitions. If a photon is absorbed by a system that possesses well-defined angular momenta, such as hydrogen atoms, the systems angular momentum quantum number changes by $\Delta\ell = \pm 1$ when it absorbs or emits a photon.

It is useful to introduce some terminology about further properties of light. When the photons also have the same phase as it is the case for a laser beam then they are *coherent*. Moreover, for a laser beam it is sometimes important that it has low *divergence*, which means that its spot size changes very little along the path of beam propagation. Light is said to be *plane polarised* if the electric field vector of all photons oscillates in parallel, i.e. in the same, fixed planes. The magnetic field vector is always polarised perpendicularly to the electric field vector, and the two vectors are in phase. A propagating wave has *circular polarisation* when the electric field vector rotates about the propagation axis and maintains a constant magnitude. The tip of the vector then describes a helix, so the light can adopt right or left circular polarisation. Plane and circular polarisation are special cases of the more general elliptical polarisation. The projection of the motion of the vector tip is then elliptical instead of circular or linear.

The wave nature of light has important consequences when we build spectrometers and study matter (Figure 2.1):

- *Light is reflected from surfaces.* If the surface is smooth down to distances below the wavelength λ the direction of reflection is everywhere the same. Mirror-like reflection where the angle of the incident and the reflected beam with the surface are the same is called specular reflection.
- *Light is scattered by the surface of powders.* Thereby, crystal faces of the small crystallites at different orientations act like mirrors which reflect the light into many different directions. This is called diffuse reflection. Light is also scattered by dust particles or by spherical droplets suspended in the air or in a liquid medium if the dimension of these particles or droplets are on the order of the wavelength of the scattered light. This is called Mie scattering. Clouds often appear white because light of all wavelengths in the visible range is scattered by small water droplets with a diameter of the order of 370–730 nm. Also milk is not transparent and appears white because of the suspended oil droplets therein. But also objects which are much

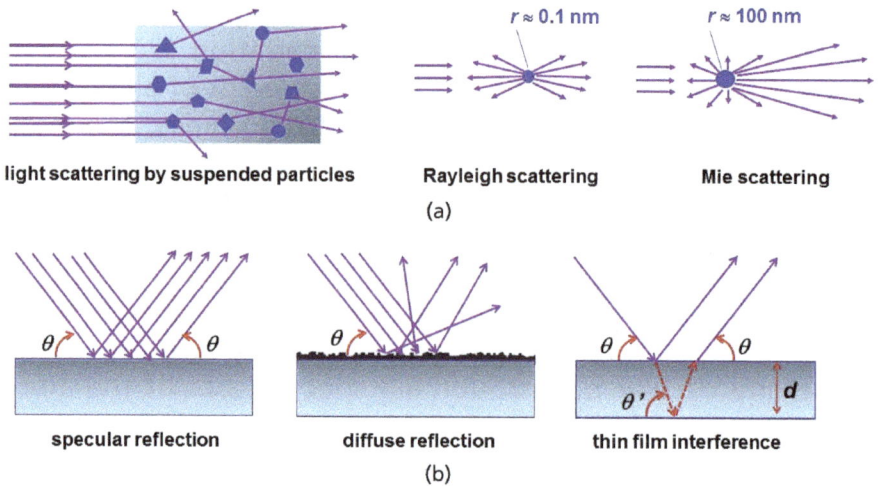

Figure 2.1:   Interaction of light with small objects and with surfaces: (a) Light is scattered by suspended dust particles or liquid droplets. Rayleigh scattering, which occurs with atoms and molecules much smaller than the wavelength of light, is symmetric in the forward/backward direction, while the forward direction is dominated for Mie scattering with spherical droplets of a size comparable to the wavelength of light. (b) Reflection on smooth surfaces is specular, on rough surfaces or on powder it is diffuse. A change in the index of refraction at a boundary causes a change of direction of the light beam, called refraction. Thin layers with a thickness d comparable to the wavelength λ reflect the incident beam from the first and also from the second surface. This second partial beam travels a longer distance than the incident beam (broken red arrows). If this distance is a multiple of 2λ this leads to constructive interference of the outgoing beam, otherwise this beam is attenuated.

smaller than the wavelength of the incident light, e.g. atoms or small molecules, scatter light. This type is called Rayleigh scattering.

- *Light may be absorbed.* If all wavelengths in the visible range are fully absorbed no light is left and the material appears black, like graphite. If only a fraction of the visible spectrum is absorbed then the other fractions are either transmitted or reflected, or both. For example, the chlorophyll molecule in green leaves of plants absorb light mostly in the red (between 650 and 700 nm) and in purple/blue part of the spectrum (between 400 and 500 nm). In between these ranges there is the green gap where relatively little absorption occurs. This is the light that we see when we look at leaves.

- *Light, just as any other type of waves, bends around an edge that it encounters*. This is called diffraction. The phenomenon is better known when light encounters a slit or hole or any object comparable in size with the wavelength of the light. The effect is amplified by periodic structures such as a grating or a crystal lattice. The incident beam of light deviates by an angle that depends on the wavelength. This effect may be used to separate the different wavelengths (colours) of sunlight into the colours of the rainbow. Diffraction of monochromatic light (light of a single wavelength) leads to typical intensity patterns due to interference.
- *Light interferes when light waves interact with each other*. When the maxima of two waves overlap, the two amplitudes add up (constructive interference); when a maximum overlaps with a minimum of another wave with the same amplitude, the two cancel each other (destructive interference). For example, when two identical waves (parallel, with identical wavelength and phase) impinge on the surface of a soap bubble with a wall thickness of the order of λ, and one is reflected from the upper surface and the other one from the lower surface the reflected beams interfere with each other. The lower beam travels a longer distance than the upper one, and in general the phase coherence will be lost in the reflected beam. However, for light of wavelength λ that encounters the surface at a specific angle such that the extra distance travelled by the second beam amounts to 2λ there will be constructive interference and the light will be seen. Therefore, the soap bubble will shine in different colours depending on the angle of incidence of the light with the bubble surface.

2.2 Absorption and Emission

The quantum nature of matter reflects the fact that individual atoms, molecules and condensed matter exist with discrete energy states. This discretisation is due to confinement of the electrons, nuclei and larger entities in the potentials given by their mutual interaction. The discrete states are ascribed to the various degrees of freedom of the electrons and nuclei, which result in electronic states of atoms, molecules and solids with delocalised electrons, to vibrational states of molecules and of solids, and to rotational states of molecules. Orbital and spin angular momenta of electrons and nuclei are further relevant degrees of freedom. Provided that there is no coupling between these degrees of freedom so that they are independent, each independent degree of freedom is characterised by a quantum number.

Matter at equilibrium assumes the lowest permitted state of energy with respect to each of the degrees of freedom, except where higher states are accessible by thermal excitation. Beyond equilibrium, transitions to higher states may be stimulated by interaction of electrical charge with the electrical field vector of electromagnetic radiation that oscillates at frequency ν (Figure 2.2(a)). The oscillating electric field polarises the molecule or particle and stimulates a transition when the photon matches the resonance condition, i.e. for $\Delta E = h\nu$.

Furthermore, certain selection rules must be obeyed. The photon may then be absorbed under promotion of the system to a higher energy state. We distinguish two cases: (i) electric dipole allowed transitions, stimulated by coupling of the electric field vector to an electron charge or to the partial charges of an electric dipole moment and (ii) magnetic dipole allowed transitions induced by interaction of a magnetic (dipole) moment with the magnetic component of the electromagnetic field.

The excited state is unstable and decays, either via stimulated emission by coupling of the electron to the oscillating electrical field, as in the absorption process, or alternatively by

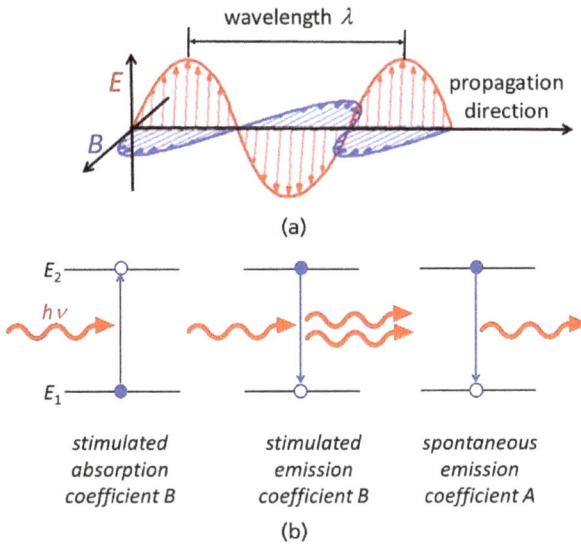

Figure 2.2: (a) Plane-polarised electromagnetic wave with oscillating electric (red) and magnetic (blue) field vectors. (b) Basic processes of absorption and emission of a photon of energy $\Delta E = E_2 - E_1 = h\nu$ between an occupied and an empty state. The coefficients A and B are the Einstein coefficients for spontaneous and stimulated (i.e. photon-induced) emission and absorption.

spontaneous emission which is independent of the alternating field (Figure 2.2(b)). Transitions can only occur between occupied initial and unoccupied final states. The electromagnetic field stimulates both, absorption and emission simultaneously so that only the net number of absorbed or emitted photons is detected in a spectroscopic experiment. The signal is then proportional to the difference in the number of excited and ground state systems, ΔN. For equal population of the two states ($\Delta N = 0$) the signal is zero, and the system is called saturated.

The stimulated signal observed for an ensemble of identical two-level systems is proportional to ΔN and to the energy density ρ of the radiation at the resonance frequency. If the light source is a thermal emitter ρ depends on temperature and is given by Planck's formula for the black-body radiation. The proportionality constant is the Einstein coefficient B that depends on the wave functions of the system (see Eq. (2.14)). For non-degenerate systems, B is identical for absorption and for stimulated emission. However, we still have the competing process of spontaneous emission, described by the Einstein coefficient A that also depends on the two wave functions but is independent of ρ. It turns out that the two coefficients are related via

$$A = \frac{8\pi h\nu^3}{c^3} \cdot B, \qquad (2.2)$$

where h is the Planck constant (6.626×10^{-34} m^2 kg s^{-1}), ν the frequency of the photons, and c the speed of light (in vacuum $c = 2.998 \times 10^8$ m s^{-1}). The two coefficients do not have the same dimension and are therefore not directly comparable. The important point in Eq. (2.2), however, is the ν^3 dependence, which means that at high frequencies or excitation energies spontaneous emission dominates over stimulated emission, whereas at low frequencies the reverse is true. We are concerned here with transitions in the UV and visible, with ν of the order of 10^{15} s^{-1} for wavelengths of 300 nm. The radiative decay of excited states will thus be predominantly by spontaneous emission, whereas in the spectral range of magnetic resonance with typical wavelengths of 3 cm for electron paramagnetic resonance (EPR) and several metres for nuclear magnetic resonance (NMR) stimulated emission prevails for typical experimental conditions.

At room temperature, the Boltzmann population of excited levels of nuclear and electron spin states is nearly equal to that of the ground state. Also, several rotational levels of molecules are reached by thermal energy, but excited vibrational states are not populated much, usually in the range of a few percent or less, and electronic degrees of freedom are not normally populated. However, there are methods of creating population inversion by high intensity pumping. This is important for laser activity, which is more easily reached for longer wavelengths due to the reduced spontaneous relaxation.

2.3 The Lambert–Beer Law and Its Limitations

Optical absorption of molecules in the gas phase or in solution is used for the determination of solute concentrations. Let us consider a beam of monochromatic light of intensity I_0 that enters a cuvette with path length d containing n absorbing molecules of absorption cross-section σ per cm^3 (Figure 2.3). The light intensity dI absorbed in a slab of thickness dx is proportional to the local intensity of light I and to both σ and n:

$$\frac{dI}{dx} = -I\sigma n. \tag{2.3}$$

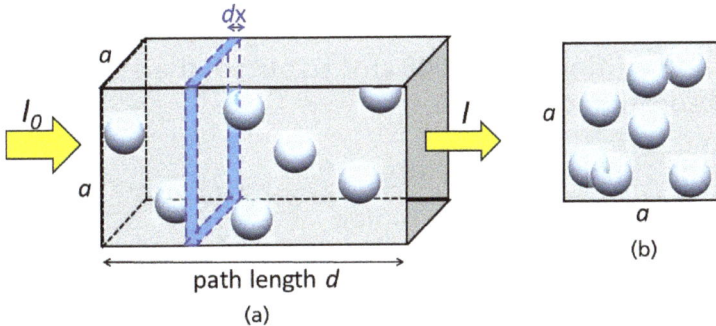

Figure 2.3: Light absorption in a cell of cross-section $a \times a$ and length d in 3-dimensional view (a) and in a projection along the beam direction (b). The intensity of the incident light is screened by the N particles with effective molecular absorption cross-section, σ. This demonstrates that the proportionality of the absorption to concentration holds only as long as the concentration is sufficiently low so that the overlap of the particles in the projection is negligible.

Separation of variables and integration, using the boundary condition $I = I_0$ for $x = 0$ leads to

$$ln\frac{I_0}{I} = \sigma nd, \tag{2.4}$$

where d is the thickness of the cuvette. Rewriting in terms of the decadic (i.e. related to log with basis 10 of I_0/I in Eq. (2.5)) molar extinction (or absorption) coefficient $\varepsilon(\lambda)$, usually given in L mol^{-1} cm^{-1}, and the concentration c in mol L^{-1} we obtain the optical density

$$log\frac{I_0}{I} = \varepsilon(\lambda)cd, \tag{2.5}$$

or in the Lambert–Beer law in its conventional form

$$I = I_0 \times 10^{-\varepsilon(\lambda)cd}. \tag{2.6}$$

$\varepsilon(\lambda)$ is normally quoted for the wavelength where the absorption has a maximum. It is characteristic for the absorbing molecule.

It is instructive to convert ε from molar to molecular units by dividing by Avogadro's number, $N_A = 6.022 \times 10^{23}$ molec mol^{-1} and writing 1 L = 1,000 cm^3 to obtain the effective molecular absorption cross-section, σ:

$$\sigma \approx \frac{2.303 \times 1000 \times \varepsilon}{6.022 \times 10^{23}} \frac{cm^3 \, mol}{mol \, cm \, molec} = \varepsilon \times 3.82 \times 10^{-21} cm^2 \, molec^{-1}$$

$$= \varepsilon \times 3.82 \times 10^{-5} Å^2 \, molec^{-1}. \tag{2.7}$$

The factor ln(10) $\approx$ 2.303 takes into account that the absorbance is normally given as log_{10} of I_0/I. Considering that large values of molar absorption coefficients of chromophores reach up to 10^4–10^5 L mol^{-1} cm^{-1} we see that for allowed transitions the effective molecular absorption cross-sections σ at a given wavelength are on the order of the geometric cross-section of the molecule.

For example, fullerene C_{60} has a diameter of 7 Å and an extinction coefficient ε(255 nm) = 175,000 L mol^{-3} cm^{-1}, which results in σ = 6.7 Å^2, while the geometric cross-section is 38.5 Å^2. In particular, larger chromophores of comparable chemical nature should be expected to be stronger absorbers. Good examples are the strongest absorption bands in the series benzene, naphthalene, and anthracene with ε of 60,000 L mol^{-3} cm^{-1} (184 nm), 133,000 L mol^{-3} cm^{-1} (221 nm), and 180,000 L mol^{-3} cm^{-1} (256 nm), respectively. They increase linearly with the number of aromatic rings. For the benzene π system, the effective value of σ amounts to 2.3 Å^2 instead of the physical cross-section of 6.2 Å^2. It should be noted that these are allowed transitions (see Section 2.4), rather than the ones of longest wavelength that are forbidden, have far lower extinction coefficients and bear no direct correlation with the geometrical cross-section.

The fact that the effective absorption cross-section at a given wavelength is in most cases smaller than the geometric value indicates that such molecules should be regarded as partially transparent. For the above example of the fullerene C_{60} we could say that only 17% of the photons which hit the molecule are absorbed. It is perhaps interesting to realise that a photon with its undoubted wave nature and a wavelength of several hundred nanometres sees a molecule much like a classical object that throws a shadow although its diameter is 1,000 times smaller than the photon wavelength. Moreover, for disc or rod shaped molecules the orientation relative to the direction of the oscillating electric field of the exciting light plays an important role (see Section 2.4).

Experimentally, for accurate determinations of ε it is important to keep the concentration of the solute sufficiently low so that the overlap in the projections of the molecules can be neglected (see Figure 2.3(b)). For benzene that has an absorption cross-section σ of 2.3 Å^2 for its strong band at 184 nm we have in the average 1 molecule in a column of cross-section σ and length 1 cm (the typical length of a cuvette) at a concentration of 6.6 μM. Since the molecules are distributed statistically, some of them may still be in the shadow of the others. 1 μM should be a safe limit for strong bands; for weaker bands the limit is correspondingly higher. At these concentrations, there should also be no interaction between the molecules that could influence the measurements. High solute concentrations can shift the band maxima or change the refractive index of the sample. The concentration dependence should be checked to detect deviations from linearity or spectral changes due to aggregation of chemical interaction. Additional errors result from light scattering due to turbid samples, or when there is partial reflection of the light from the cuvette (see also Chapter 3). The latter effect is minimised by using double beam spectrometers and a reference cuvette with an identical cuvette containing the solvent.

2.4 Transition Moment and Selection Rules

2.4.1 *The transition dipole moment*

The probability that a system undergoes an electric dipole allowed transition from the ground state to an excited state under the influence of an electric field that oscillates on resonance is calculated from the expectation value of the electric dipole operator $\mu = -er$, e is the electrical charge and r is the position vector of the electron, with components x, y and z. The transition dipole moment μ_{fi} is calculated for the electronic wave functions of the initial and the final states, ψ_i and ψ_f.

$$\mu_{fi} = \int \psi_f^* \mu \psi_i d\tau = -e \int \psi_f^* r \psi_i d\tau. \tag{2.8}$$

For $i = f$, i.e. when the final state is the same as the initial state, this equation provides the permanent electric dipole moment of the system. For $i \neq f$ the integral provides the transition dipole moment, which is a measure of the probability that a transition from the initial to the final state may be stimulated by the electric field oscillating at a frequency that corresponds to the resonance condition, $\nu = (E_f - E_i)/h$. Since the molecular dimension is implicit in r it is qualitatively conceivable without calculation from Eq. (2.8) that μ_{fi} increases for larger molecules. The transition dipole moment is measured in the same units as the electric dipole moment of the ground state molecule, i.e. in Debye (1 D = 3.336×10^{-30} C m).

Selection rules derive from the expression for the transition moment. In the absence of symmetry (excluding the trivial identity operation) the integral of Eq. (2.8) is non-zero, and all transitions are allowed. However, for highly symmetric wave functions as in free atoms or molecules in isotropic environment, the integral can be zero based on symmetry alone. The relevant criterion is the presence of a centre of inversion i. Orbitals, or functions in general can be symmetric (gerade, g, German for even) or antisymmetric (ungerade, u, for odd) with respect to i, depending on whether the function retains or changes its sign on inversion at i. The classifications ungerade (u) and gerade (g) are explained in Figure 2.4. (a) and (b) are simple mathematical functions $f(x)$. For (a), $f(x)$ is proportional to $\exp(-x^2)$, and $f(x) = +f(-x)$, meaning that $f(x)$ is gerade with respect to the origin i of the coordinate system. The integral over x, i.e. the area under the curve, is non-zero. For (b), $f(x) = -f(-x)$, meaning that $f(x)$ is ungerade. It is seen immediately that the negative and the positive areas cancel so that the integral is zero. (c) is a vector along x, like the position vector r. It has a positive and a negative end and is always *ungerade*. If the wave functions ψ_i and ψ_f are of the same parity, i.e. both g or both u, then the product $\psi_f^* \, r \, \psi_i$ has the parity u and the integral is zero. This is equivalent of saying that the transition between the two states is forbidden.

2.4.2 *Selection rules due to orbital symmetry*

Free atoms and ions are centrosymmetric, their s and d orbitals are of g parity, p and f orbitals are of u parity. Therefore, s $\leftrightarrow$ s,

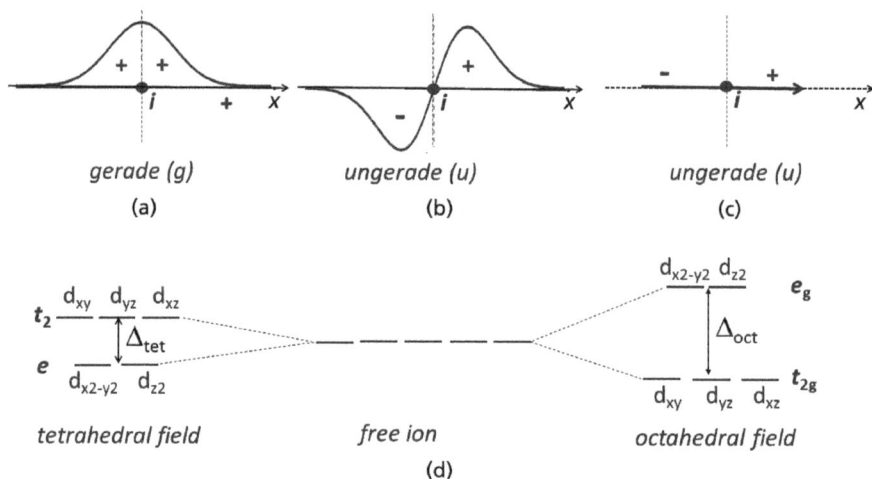

Figure 2.4: Parity of mathematical functions: (a) $f(x)$ is proportional to $\exp(-x^2)$ is of g parity. (b) $f(x)$ is proportional to $x \cdot \exp(-x^2)$, which is of u parity. (c) A vector is always u. Function (b) is the product of the functions (a) and (c). (d) Energy level diagram of transition metal d orbitals as free ion and in tetrahedral and octahedral crystal field.

s $\leftrightarrow$ d, p $\leftrightarrow$ f transitions are forbidden, whereas s $\leftrightarrow$ p, p $\leftrightarrow$ d and d $\leftrightarrow$ f are allowed in both directions. Based on parity alone, s $\leftrightarrow$ d should also be allowed, but it is in fact not since one of the selection rules for hydrogen-like atoms is $\Delta\ell = \pm 1$. Parity is thus a convenient but not a sufficient criterion to judge transitions. This is seen also in spectra of transition metal ions in ligand or crystal fields, where optical spectroscopy is well established. For such systems of centrosymmetric symmetry, the Laporte rule states in agreement with what was said previously that transitions between orbitals of the same parity are forbidden. Transitions within a given atomic subshell, e.g. between a set of p or of d orbitals should therefore not be allowed. To first order, they are degenerate in energy anyway, but in a crystalline or molecular environment the centrosymmetry is broken (Figure 2.4(d)). In an octahedral field the degeneracy is lifted into two sets of energy levels, but the centre of symmetry is retained in an octahedron so that all orbitals are still of g parity as in the undistorted case. This means that transitions between the triply degenerate t_g and the doubly degenerate e_g are very weak and only allowed by coupling to vibrations. However, in an environment that imposes a tetrahedral field the centre of symmetry is not retained, and parity of the orbitals is no longer a good criterion. d $\leftrightarrow$ d transitions are therefore of higher intensity in tetrahedral than in octahedral environments. We learn that selection rules are often not so rigid, and transitions that are forbidden in first order are often weakly allowed if symmetry breaking effects like crystal fields, vibrations or collisions are present.

2.4.3 *Selection rules due to spin functions*

Spectroscopic transitions that are induced by the electrical component of the electromagnetic field are electric dipole allowed transitions. These are the electronic transitions in the UV or the visible range of the electromagnetic spectrum, vibrational transitions in the infrared, and rotational transitions in the microwave range. For each of these transitions, the integral of Eq. (2.8) is evaluated with the corresponding wave functions, i.e. with the electronic, vibrational or rotational functions which all have separate sets of quantum numbers and selection rules. Apart from these, there are also magnetic dipole allowed transitions, induced by the magnetic component of the electromagnetic field. This component acts on the magnetic part of the wave function, i.e. on the electron and the nuclear spins. The selection rules that govern these transitions are much more rigid. The spin state is thus in most cases retained, and singlet remains singlet, triplet remains triplet. Intensities of absorptions which violate spin selection rules are therefore very weak and not normally observed. An exception is the radiative deactivation of triplet states, called phosphorescence (see Figure 2.9). This is reflected in the much longer radiative decay time of triplet than of singlet states, as seen by comparison of phosphorescence and fluorescence lifetimes. Transitions between different spin states are facilitated by spin–orbit coupling.

2.4.4 Raman spectroscopy

Raman spectroscopy also uses interaction of matter with light, typically with lasers in the visible range. It is based on monochromatic excitation mostly outside the resonance condition into virtual states (non-eigenstates). The exciting photon is not absorbed, only scattered. A small fraction of the scattered photons are of a higher or lower energy than the incoming radiation; the difference corresponds to energy obtained by deactivation or excitation of vibrational and rotational states of molecules or condensed matter. These transitions are induced by polarisation of the matter in the exciting oscillating electrical field, as in absorption spectroscopy. The dipole operator in Eq. (2.8) is then given by $\mu = \alpha \cdot E$, with α being the molecular polarisability tensor and E the exciting electric field. These off-resonance interactions produce spectra with weaker intensities than absorption spectra by several orders of magnitude. Raman spectroscopy is therefore of much lower sensitivity. It is mostly used to observe molecular vibrations and will not be discussed further here.

2.4.5 The orientation of the transition dipole moment

$|\psi_i|^2$ describes the electron density distribution in the initial orbital, $|\psi_f|^2$ in the final state. Thus, the electronic transition moment describes a redistribution of electron density along the direction of the oscillating electric field. This is illustrated in Figure 2.5 for the transition of the hydrogen 1s electron to its $2p_z$ state. The binding energy of the initial state is the negative ionisation potential of H that amounts to -13.6 eV, while that of the 2p state is -3.4 eV. The excitation energy thus corresponds to 10.2 eV or a wavelength of 121.6 nm.

A pure electronic transition without any change in vibrational or rotational excitation is a nearly monochromatic sharp

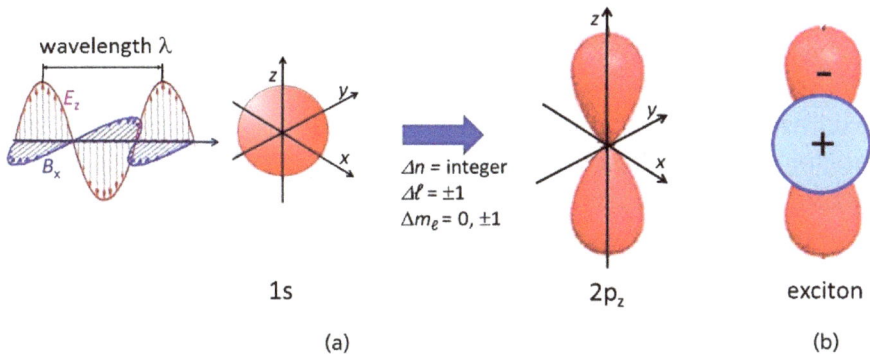

Δn = integer
$\Delta \ell = \pm 1$
$\Delta m_\ell = 0, \pm 1$

1s 2p$_z$ exciton

(a) (b)

Figure 2.5: (a) The z-component of an oscillating electric field couples to the electron charge in the 1s orbital of a hydrogen atom (initial state). When the frequency corresponds to the resonance condition, this stimulates an electronic transition to the $2p_z$ final state, which represents a redistribution of electron density. (b) The difference between the final and the initial state is called an exciton. It represents a bound state between the promoted electron (red) and the hole that is left behind in the initial 1s state (blue).

spectral feature. However, the same transition occurs with simultaneous excitation of vibrational and rotational states, which leads to slightly higher frequencies (see Section 2.6) and produces a multitude of unresolved or partly resolved features, especially under line broadening conditions in solution or in a pressurised gas where rotational states are not well defined. Furthermore, rotational and, to a lesser extent, also vibrational states are thermally excited in the electronic ground state. They can de-excite during electronic excitation, which leads to transitions at lower frequencies. Both cases represent the same electronic transition, and the total intensity or band area is given by integration of the band profile $\int \varepsilon(\nu)d\nu$. It can be shown that the total absorption intensity or the band area of an electronic transition is proportional to $|\mu_{fi}|^2$.

Figure 2.5 is also convenient to explain the term exciton that is widely used for semiconductors, insulators, and also for some liquids. It stands for a neutral quasiparticle, a bound state between a negative electron and a positively charged electron hole that is left behind in the initial orbital when an electron is excited to the final orbital. The exciton is conveniently regarded as the difference between the electron density after and before excitation. By taking this difference, the stationary part, i.e. the nuclei, the atomic core orbitals and the non-involved valence orbitals are subtracted out. What is left is the exciton, which is often mobile and can transport excitation energy in solids. For the excitation of the hydrogen 1s to $2p_z$ state the exciton is shown in Figure 2.5(b). Its binding energy amounts to -3.4 eV, the energy of the $2p_z$ state.

The electromagnetic field oscillates in time and an electric field polarised along the z-axis (such as in Figure 2.5) accordingly polarises the 1s orbital of H alternatingly in the positive and negative z directions. This time dependence is not taken into account in Eq. (2.8). Although the transition moment μ_{fi} has a direction and a magnitude and is normally termed a vector, its sign is arbitrary and only its squared value can be experimentally observed. Furthermore, the hydrogen-like orbitals in Figure 2.5 reflect the time-averaged electron density distribution although the quantum mechanical calculation of these orbitals treats the electron as a point-charge with explicit coordinates relative to the nucleus. The wave functions are derived using the time-independent Schrödinger equation which does not contain time as a variable. If one wanted to calculate the trajectory of the electron motion one would need to know the initial conditions of its motion, i.e. its initial coordinates and momentum. For many-electron systems this would lead to an immense increase in complexity that is avoided by using the time-independent Schrödinger equation.

The free hydrogen atom is isotropic. The direction of the exciting electrical field vector is therefore irrelevant, and an electron can always be excited provided that the resonance condition of the radiation is met and the integral of Eq. (2.8) is not zero. If the integral is equal to zero, the transition is said to be forbidden, otherwise it is allowed with an intensity depending on the value of the integral. Evaluation of this integral for the H atom using the analytical expressions for the wave functions reveals that there is no constraint on the principal quantum number n, i.e. Δn can take any integer value. However, the orbital angular momentum quantum number must change as $\Delta \ell \pm 1$, which accounts for the fact that the angular momentum must be conserved when the photon of circularly polarised light is absorbed or emitted during the transition. A further constraint for a non-zero integral is that the z-component of the orbital angular momentum changes as $\Delta m_\ell = 0, \pm 1$. These constraints on the change of the various quantum numbers are called selection rules.

2.4.6 *The case of benzene*

It is instructive to discuss the electronic transitions of benzene on the basis of the Hückel molecular orbital scheme (Figure 2.6(a)). In the framework of the linear combination of atomic orbitals to form molecular orbitals (LCAO–MO), the six p-type AOs perpendicular to the molecular plane combine to six π-type MOs whose energies increase with the number of nodal planes where the sign of the wave function changes. There are two degenerate π HOMOs and two degenerate π^* LUMOs (degenerate means they have equal energy). The red arrows mark the possible transitions of lowest energy or longest wavelength. In this energy diagram the levels represent single electron energies, as commonly used in chemistry, and the four transitions should be expected to be of equal energy and appear as indistinguishable features in the spectrum. This picture is inaccurate by a large amount since due to the electron interaction the energy of an orbital depends on its occupancy. For example, on the basis of the Rydberg formula for hydrogen-like atoms the 1s electrons of He should have an energy of −54.4 eV. The ionisation potential of He$^+$ (i.e. the species where only a single electron is present) is indeed 54.4 eV,

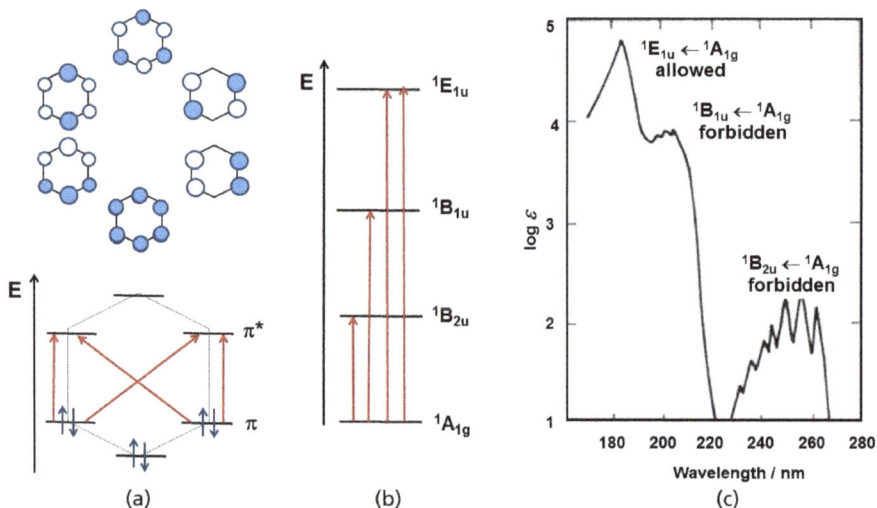

Figure 2.6: (a) π-type molecular orbitals of benzene with corresponding energy level diagram with transitions between the highest occupied πMO (HOMO) and the lowest unoccupied π^* MO (LUMO). The size of the circles represents the squared coefficient (weighting factor) of the AO in the linear combination to the MO, and empty/filled circles indicate negative/positive signs of the wave function. (b) Energy diagram of electronic transitions of benzene, and (c) $\pi^* \leftarrow \pi$ electronic transitions within the singlet manifold of benzene (adapted with permission from Ref. [2], © (1961) American Institute of Physics).

but the first ionisation potential of He^0 is 24.6 eV. The difference of 29.8 eV is due to the electron–electron repulsion (also called the electron correlation energy). The benzene molecule is much larger, therefore the electrons can avoid each other much better and are at a larger distance on average. The electron correlation energy for an electron–hole pair at a distance of 0.56 nm, the diameter of the benzene molecule, is 2.56 eV, which is far from negligible. Depending on the density distribution of the electron in the LUMO and that of the hole in the HOMO from which the electron was excited, and taking into account that there are two different options for the HOMO and two for the

LUMO, there is a very noticeable difference in the interaction. Because a quantitative calculation of electron correlation is not easy, the situation is far from trivial. The assignment of the spectrum shown in Figure 2.6(c) has therefore stimulated a lot of discussions, but there is now agreement that this HOMO–LUMO transition gives rise to the three observed bands since two of the transitions are degenerate. The point group of benzene is D_{6h} in the Schönflies notation, representing the 6-fold symmetry of the planar molecule. The states and the transitions are labelled by symbols of this point group. We cannot go into group theoretical details here, but in short, the symbol A represents a fully symmetric state (the ground state of molecules is generally of A character), B states are not fully symmetric, and E states are doubly degenerate. The superscript "1" denotes that all states considered here are singlet states, and the subscripts g and u reflect the character towards inversion at a centre (Figure 2.4). The subscript is amended by a number to distinguish between states of otherwise identical symmetry.

Next, we look at the intensities of the transitions in Figure 2.6(c), remembering the logarithmic scale of the extinction coefficient ε. The band at 184 nm has $\varepsilon = 60,000$ L mol^{-1} cm^{-1}, and the transition is doubly degenerate and allowed, as already discussed in Section 2.3. The band near 200 nm is forbidden and has $\varepsilon = 8,000$ L mol^{-1} cm^{-1}, an order of magnitude less than the strong transition. The best known transition of benzene is its vibrational multiplet around 255 nm. It is also forbidden and actually with $\varepsilon = 215$ L mol^{-1} cm^{-1} quite weak. Why then are symmetry-forbidden transitions observed? The answer lies in the limited rigidity of symmetry. The symmetry of the benzene molecule is D_{6h}, but this is the average structure. In reality, the molecule vibrates, so that its true structure is vibrationally distorted at almost any instant. Vibrations and also distortions as a consequence of collisions are common causes of violations of selection rules. They play an important role also for transitions by internal conversion.[3,4]

2.4.7 *The case of naphthalene*

Figure 2.7(a) shows the HOMO and the LUMO+1 π orbitals of naphthalene. The *g–u* parity of molecular orbitals is judged by the symmetry of the wave function with respect to the

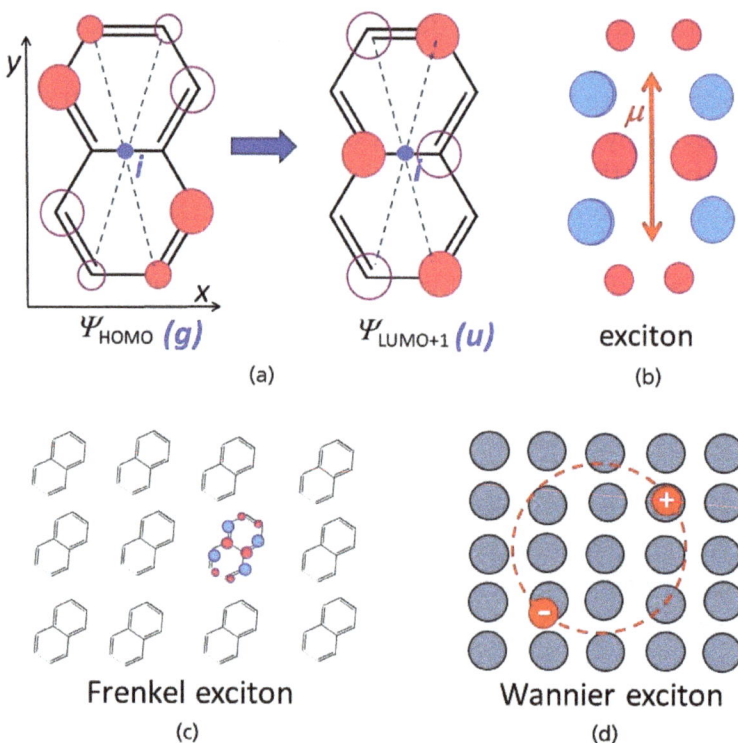

Ψ_{HOMO} *(g)* Ψ_{LUMO+1} *(u)*   exciton

(a)   (b)

Frenkel exciton   Wannier exciton

(c)   (d)

Figure 2.7:   (a) The HOMO to LUMO+1 transition of naphthalene is polarised along the long axis of the molecule. The size of the pink circles reflects the squared molecular orbital coefficients, i.e. the one-electron density distribution within a given orbital, giving an impression of the change during the transition. Open and filled circles represent opposite signs of the wave functions. (b) The exciton reflects the difference between the one-electron distributions of the two orbitals, the blue circles represent the hole. The transition moment μ is represented by a double arrow (red), although it is normally termed a vector, its sign is arbitrary. (c) is an image of the localised Frenkel exciton in the naphthalene crystal. (d) shows schematically the more extended Wannier exciton in an inorganic semiconductor.

molecular inversion centre i. The filled and empty pink circles represent the opposite signs of the wave function, and the area of the circles is the square of the coefficient of the atomic p orbital, i.e. the single electron density distribution. It is obvious that ψ_{HOMO} is symmetric (g) with respect to inversion at i, while ψ_{LUMO+1} is antisymmetric (u). The opposite parity means that the integral of the transition moment is non-zero and the transition between the two orbitals is allowed. This is a measure of the magnitude of the transition moment μ_{fi} but it says nothing about its direction. In order to find the direction we have to evaluate the components from $\mathbf{r} = \mathbf{x} + \mathbf{y} + \mathbf{z}$. Since the molecule is planar, the z component perpendicular to the molecular plane is zero. The y component along the long molecular axis of ψ_{HOMO} is antisymmetric, or of u parity, while the corresponding component of ψ_{LUMO+1} is of g parity. The overall parity is therefore $u \cdot u \cdot g = g$, giving a non-zero integral. The x component is evaluated in an analogous way, which yields a total parity of $u \cdot u \cdot u = u$, revealing that this component of the transition moment is zero. We therefore conclude that the electric field vector of the exciting radiation, or at least a component of it, must oscillate parallel to the long axis of naphthalene in order to stimulate a transition between the two chosen orbitals (Figure 2.7(b)). This is consistent with the symmetry of the molecule, which demands that the transition moment can only be either parallel to the long or to the short axis.

2.4.8 *Excitons*

Figure 2.7(b) shows the redistribution of the one-electron density in the two orbitals due to the transition between the HOMO and the LUMO + 1 orbitals. In particular, it shows the structure of the exciton, the bound state between electron and hole in which the backbone of the molecule has been subtracted out. There is a pronounced modulation of the electron–hole density distribution over the molecule, but the dipole moment of the excited state remains zero. Nevertheless, the dimension of the transition moment is the same as that of an electric dipole moment. The symmetry of the exciton is equal to that of the ground state.

Excitation of an organic molecule in a solid is an event that is localised because the wave functions are localised to the molecule. The excitation energy in the solid is therefore similar to that of the isolated molecule. The exciton can hop from one molecule to the next, providing a mechanism to transport energy in a crystalline solid. An electron and hole remain mainly localised on the same molecule when the exciton migrates. These localised electron–hole pairs are called Frenkel excitons (Figure 2.7(c)). In contrast, the ground state wave functions of valence electrons in semiconducting solids are delocalised over the entire crystallite. This is not the case for an exciton whose electron and hole remain bound because of their Coulomb attraction. The binding energy is much less than in molecules, typically only ca. 10 meV. This fact is reflected in slight shifts of the energy levels of the electron and hole defect states with respect to the band edges of the unperturbed solid into the band gap. The wave functions of these electron–hole pairs extend over many atoms (Figure 2.7(d)) and are to some extent modulated by the presence of these atoms. Nevertheless, they are described well by positronium-like wave functions with a much larger Bohr radius of typically 10 nm. Positronium, Ps, is an exotic hydrogen-like atom consisting of a positron and an electron. The Coulomb attraction is shielded by the host matter and reduced by the dielectric constant of the solid. These non-localised excitons in semiconductors are called Wannier excitons. There is also a variant of the Frenkel exciton in which electron and hole are localised on neighbouring molecules in the crystal. These are called charge transfer excitons.

Due to the interactions of molecules in the lattice of a solid, their absorption bands may be shifted, broadened or split relative to what is observed in a dilute solution. We distinguish two

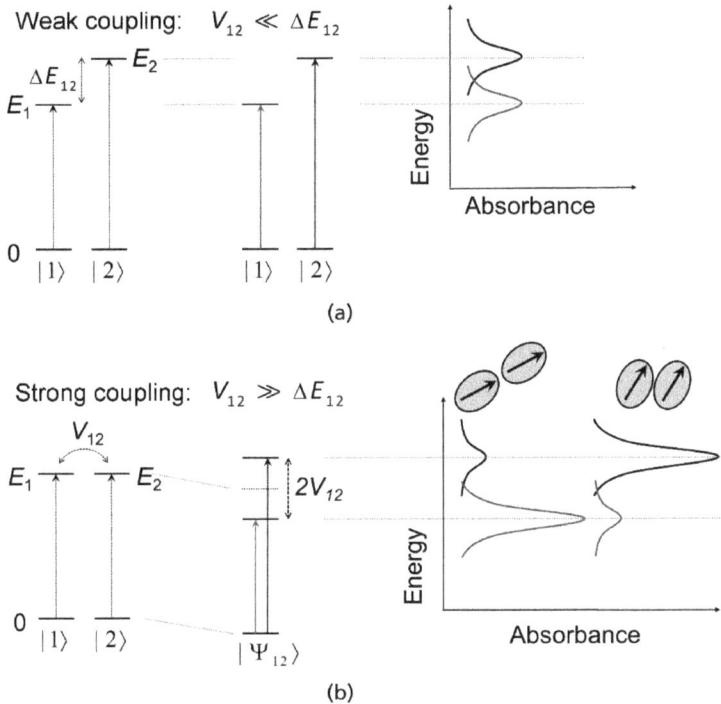

Weak coupling: $V_{12} \ll \Delta E_{12}$

(a)

Strong coupling: $V_{12} \gg \Delta E_{12}$

(b)

Figure 2.8: Effects of the interaction (V_{12}) and site-energy difference $\Delta E_{12} = |E_1 - E_2|$ on the energy levels and spectra of two monomer chromophores with states $|1\rangle$ and $|2\rangle$. (a) In the weak-coupling limit the interaction between the chromophores is weak compared to their site-energy difference and the resulting absorption spectrum is independent of the mutual orientation of the chromophores. (b) In the strong-coupling limit the difference in site energies is negligible and two new, delocalised exciton states are created. Spectra are shown for the two dipoles arranged in a "head-to-tail" and a "sandwich" configuration.

causes of this and discuss it for the example of an interacting pair of chromophores (Figure 2.8). First, the two molecules of the pair may have different excitation energies because they either occupy inequivalent (not symmetry-related) sites in the unit cells, or they may experience static disorder in an environment that is not crystalline, as in biological environments. This is the case of site energy splitting but weak electronic coupling, shown in Figure 2.8(a). The effect resembles the solvatochromic effect that will be discussed in Section 2.8. Secondly, a Frenkel exciton is not always localised on a single molecule. If two chromophores are sufficiently close to each other such that their wave functions overlap, excitonic coupling occurs between the two chromophores. In such a case, the wave function of the excited state is delocalised over both molecules and can be described as a linear combination of the individual molecular wave functions $|1\rangle$ and $|2\rangle$ to the admixed state $|\Psi_{12}\rangle$ in a similar way as for LCAO theory. For such an excitonically coupled dimer, Coulombic interactions occur between the electrons and nuclei of one molecule with those of the other one. For uncharged molecules, this interaction can be approximated by a dipole–dipole interaction. As a result of this interaction, the excited state is split into two excitonic states with distinct energies such that the energy split equals twice the coupling energy, a phenomenon known as Davydov splitting.[5,6] This excitonic splitting for the strongly coupled dimer is illustrated in Figure 2.8(b). When the two molecules are in a "head-to-tail" configuration, the lowest energy transition is strongly allowed (it borrows intensity from the positively admixed wave function of the second molecule in the pair by increasing the transition dipole moment) and the excitonic state is redshifted, while a blueshift occurs for a "sandwich" dimer. This effect is of key importance in natural photosynthesis where extremely high chromophore densities are present in the light-harvesting complexes. As a result, the absorption spectrum of a photosynthetic light-harvesting complex is very sensitive to the relative positions and orientations of the chromophores (see Section 7.5).[7,8]

When the wave functions of the two chromophores do not overlap or when their interaction with the vibrational modes of the environment (the phonon bath) dominates their interaction with each other, the Frenkel exciton will be localised to each of the individual chromophores and their mutual orientation does not affect their absorption spectra (Figure 2.8(a)).

2.5 Oscillator Strength

The extinction coefficient $\varepsilon(\lambda)$ that is derived from the Lambert–Beer law describes the factor by which light transmitted through a medium is diminished at a given concentration. It is normally quoted for the maximum of the absorption band of a transition, λ_{max}, but since the band extends over a variable range of frequencies the value of $\varepsilon(\lambda_{max})$ is of limited value to characterise a transition. A more rigid quantity is the electric dipole oscillator strength f that relates to the area A, a dimensionless integrated extinction coefficient:

$$f = 4\ m_e c\, \varepsilon_0 A\, /\, N_A e^2 = 4.33 \times 10^{-9} \int \varepsilon(\tilde{\nu}) d\tilde{\nu}, \qquad (2.9)$$

where m_e is the electron mass, c the velocity of light, ε_0 the vacuum permittivity, N_A Avogadro's number, e the electron charge, and the integration over the absorption band is in units of wave numbers.

Besides the experimental determination of Eq. (2.9), f is defined from theory as:

$$f = (4m_e h\nu B)\, /\, e^2 = 8\pi^2 m_e \nu\, |\mu|^2\, /\, 3e^2 h, \qquad (2.10)$$

where $B = 4\pi^2 |\mu|^2 / 6\varepsilon_0 h$ is the Einstein coefficient of stimulated absorption, h is Planck's constant and $|\mu|$ is the transition dipole moment (Eq. (2.8)). f is a measure of the strength of a transition, the ratio of the actual intensity and the intensity radiated by an electron oscillating harmonically in three

Table 2.1: Values for extinction coefficients and oscillator strengths to typical electronic transitions between singlet (S) and triplet (T) states.

Molecule	Transition	$\tilde{\nu}_{max}$ (cm^{-1})	ε_{max} (L mol^{-1} cm^{-1})	f	k_e (s^{-1})
p-Terphenyl	$S_1(\pi, \pi^*) \to S_0{}^a$	30,000	3×10^4	1	10^9
Perylene	$S_1(\pi, \pi^*) \to S_0{}^a$	22,850	4×10^4	0.1	10^8
p-Xylene	$S_1(\pi, \pi^*) \to S_0{}^a$	36,000	700	0.01	10^7
Pyrene	$S_1(\pi, \pi^*) \to S_0{}^a$	26,850	500	10^{-3}	10^6
Acetone	$S_1(n, \pi^*) \to S_0{}^a$	30,000	10	10^{-4}	10^5
Xanthone	$T_1(n, \pi^*) \to S_0{}^b$	15,000	1	10^{-5}	10^4
Acetone	$T_1(n, \pi^*) \to S_0{}^b$	27,000	0.1	10^{-6}	10^3
1-Bromo-naphthalene	$T_1(\pi, \pi^*) \to S_0{}^b$	20,000	0.01	10^{-7}	100
1-Chloro-naphthalene	$T_1(\pi, \pi^*) \to S_0{}^b$	20,000	10^{-3}	10^{-8}	10
Naphthalene	$T_1(\pi, \pi^*) \to S_0{}^b$	21,300	10^{-4}	10^{-9}	0.1

Notes: [a]Spin allowed transition, [b]spin forbidden transition.

dimensions.[9] For such an ideal oscillator the strength is unity. Experimentally, values near unity are found for strongly allowed electronic transitions in molecules. Typical values are listed in Table 2.1.

2.6 The Jablonski Diagram of Molecular Chromophores

Transitions between electronic states of molecules are best discussed in using the Jablonski energy diagram (see Figure 2.9). Most molecules possess a singlet ground state S_0 (no unpaired electrons), and absorption of a photon corresponding with an electron dipole allowed transition promotes an electron from S_0 to a singlet excited state S_1, S_2 or higher. Direct transitions into a state of different spin multiplicity (e.g. T_1) are spin-forbidden. The electronic states are to a good approximation pure spin states, with wave functions uncontaminated by other spin states, characterised by good quantum numbers. Transitions to

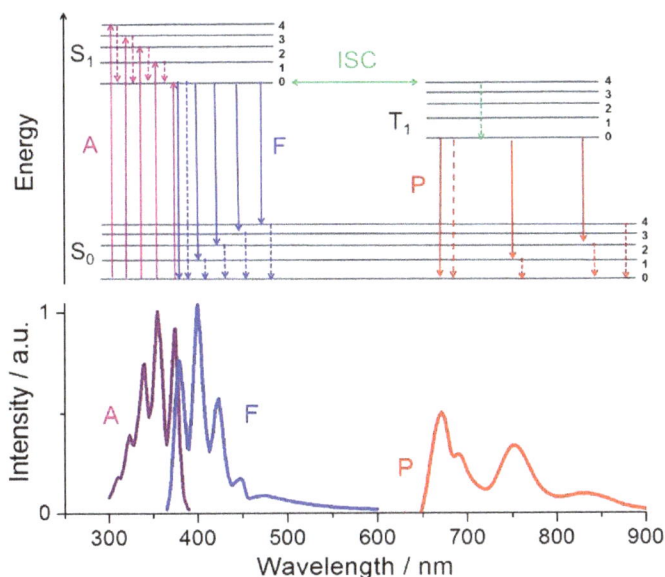

Figure 2.9: Jablonski energy diagram of electronic transitions, based on data for anthracene: Absorption A from ground state S_0 to excited state S_1 and higher singlet states (time scale of 10^{-15} s), fluorescence F from S_1 to S_0 (typically in 0.1–30 ns for organic chromophores, see Table 2.2), phosphorescence P from excited triplet state T_1 to S_0 (10^{-3} to 10^{+2} s). These are all radiative processes and shown with full line arrows. Broken line arrows indicate non-radiative transitions within one manifold, called vibrational relaxation, or between two manifolds, called internal conversion (IC). The fine structure in A, F, and P is due to vibrational sublevels of S_1 (for A) and of S_0 (for F and P), giving rise to characteristic mirror image type spectra for A and F. It should be noted that the y-scale of the upper part of the figure is the energy E, while the x-axis of the lower part is given as the wavelength, which is related to the energy difference ΔE through $\lambda = hc/\Delta E$. This inverse relationship leads to the pronounced apparent increase in spacing of vibrational lines at large values of λ.

states of different spin multiplicity, called intersystem crossing (ISC), are induced by spin–orbit interaction, which is a heavy-atom (relativistic) effect. Thus, the presence of heavy atoms such as iodine substituents of the molecule, or collisions with a heavy noble gas atom (e.g. Xe) or a heavy ion in solution promote ISC.

As discussed in Section 2.2, the photon frequency plays an important role in a transition. The time-dependent Schrödinger equation determines that the time-evolution between two states occurs with a frequency given by $v = \Delta E/h$, where ΔE is the energy difference between the involved states and h is the Planck constant. Therefore, stimulated electronic transitions are extremely fast and occur at the frequency of the UV/vis spectrum, i.e. on the order of 10^{-15} s. This means that optical spectroscopy is extremely fast, faster than the nuclear motion of vibrations, and it captures an instant picture of the observed object. Fluorescence, however, is a spontaneous process that is slower at these frequencies by several orders of magnitudes. Phosphorescence, also a radiative process, is slowed down because it is spin forbidden and can be promoted by the same mechanisms as discussed for ISC.

Because of the low phosphorescence yields and their long lifetimes, triplet states are sometimes called *dark states.* A rule-breaking exception was recently reported for caesium lead halide ($CsPbX_3$, where X = chloride, bromide or iodide) of the perovskite structure. This is a cubic structure (the name-giving structure is actually $CaTiO_3$ with orthogonal lattice distortion) consisting of corner-shared octahedrons with Pb^{2+} in the centre surrounded by the large halide ions. The Cs ions occupy the gaps between the octahedrons. The perovskites emit light much more efficiently than expected, because of the strong spin–orbit coupling imposed by the various heavy elements.[10] It couples singlet and triplet spin moments with orbital moments to total spins $J = 0, \pm1$. In these cases, fluorescence and phosphorescence do not have the strict conventional meaning, and we simply talk of photoluminescence.

Vibrational transitions occur in the infrared and their energy is dissipated as heat. In Figure 2.9, five vibrational sublevels are drawn and numbered for each electronic state. Their spacing decreases with increasing energy, because of bond anharmonicity. In principle, there are also rotational sublevels which are, however, not normally resolved in condensed phase spectroscopy, although they also play a role in heat dissipation of non-radiative deactivation.

Fluorescence and phosphorescence are in competition with the non-radiative deactivation channels between electronic states. Vibrational relaxation occurs on a typical time scale of 10^{-14}–10^{-11} s, and ISC between S_1 and T_1 at 10^{-8}–10^{-3} s. ISC may be reversible, depending on the relative energy of the states and on the nature of the molecules. Fluorescence is observed when it is faster than or on the same order as radiationless deactivation plus ISC. Conversely, for phosphorescence to occur ISC must be faster than or on the same order as fluorescence plus radiationless deactivation of S_1, and in addition, phosphorescence emission must be faster than or on the same order as radiationless deactivation of T_1 that occurs typically in 10^{-6} –10^{-2} s, making phosphorescence a rarely observed process at room temperature.[11] The latter condition is much harder to meet than the condition for fluorescence, and the spin-forbidden phosphorescence is therefore much less common than fluorescence. Vibronic interactions facilitate the non-radiative transitions, but at cryogenic temperatures (77 K or even <10 K) these can be "frozen out" so that luminescence experiments are often more promising at low temperature.[11]

Fluorescence occurs from the vibrational ground state ($v' = 0$) of the excited singlet state S_1. Because of the preceding vibrational relaxation fluorescence is of longer wavelength and of lower energy than absorption. For its detection, excitation is normally done at a fixed wavelength, conventionally at the maximum of the corresponding absorption band, although the fluorescence quantum yield (the fraction of fluorescence photons relative to the number of absorbed photons) is often independent of the excitation wavelength (Kasha's rule). Emission occurs in all spatial directions, although not with equal probabilities, and in order to minimise interference by the scattered incoming light fluorescence is experimentally often detected at an angle perpendicular to the incoming beam of light (see Figure 3.3). In such a geometry, a large fraction of the emitted photons misses the detector. This can be avoided when the photons are collected in all spatial directions using an integrating Ulbricht sphere.

If the exciting light is plane polarised (i.e. if the exciting electric field vectors of all photons oscillate in the same plane) and if there is only little rotational diffusion of the chromophore during the fluorescence lifetime so that the transition moment that is coupled to the polarisation of the exciting light keeps its orientation, the fluorescence will partly retain this polarisation P. The latter is defined as

$$P = \frac{I_\| - GI_\perp}{I_\| + GI_\perp},$$
(2.11)

and the anisotropy A as

$$A = \frac{I_\| - GI_\perp}{I_\| + 2GI_\perp},$$
(2.12)

where $I_\|$ and $I_\perp$ are the fluorescence intensities parallel and perpendicular to the polarisation direction of the exciting light. G is an instrumental factor that is unity if the detector sensitivities for parallel and perpendicular radiation are the same. The denominator in Eq. (2.12) is the total fluorescence intensity. Fluorescence depolarisation measurements provide a powerful means for studying the dynamics of the dye molecules in their environment.

Some standard values for the fluorescence lifetime are given in Table 2.2. They are independent of excitation and emission wavelength and only weakly solvent-dependent. By far the shortest lifetime is found for the iodine-substituted erythrosine dye. This is due to the heavy-atom effect (spin–orbit coupling) of iodine which enhances the ISC rate by admixing triplet character to S_1.

The T_1 state is of lower energy than S_1 by the so-called singlet–triplet splitting ΔE_{ST} of usually between 50 meV to >1 eV.[11] Both excited electronic configurations have one electron in their former HOMO and one in their LUMO orbital. According to

Table 2.2: Standard values for fluorescence lifetimes, measured in liquid solution at 20°C.[12]

Molecule	Solvent	Lifetime[a] (ns)	λ_{ex} (nm)	λ_{em} (nm)
Anthracene	Methanol	5.1 (3)	295–360	375–442
	Cyclohexane	5.3 (1)	295–360	375–442
9-Cyanoanthracene	Methanol	16 (1)	295–360	400–480
	Cyclohexane	12.7 (7)	295–360	400–480
9,10-Diphenylanthracene	Methanol	8.7 (5)	295–360	400–475
	Cyclohexane	7.5 (4)	295–360	400–475
Rhodamine B	Water	1.74 (2)	488–575	560–630
	Methanol	2.5 (1)	488–568	550–630
N-(3-Sulfonylpropyl) acridinium	Water	31.2 (4)	300–330	466–520
Erythrosine B[b]	Water	0.089 (3)	488–568	550–580
	Methanol	0.47 (2)	488–568	550–590

Notes: [a]Error of last digit in parentheses, [b]Iodine-containing compound.

Hund's rule the configuration with the highest spin multiplicity has the lowest energy. ΔE_{ST} lies typically between 50 meV and >1 eV.[11] Phosphorescence is therefore well separated from fluorescence in the luminescence spectrum, as seen in Figure 2.9. It can be recorded separately using a long-pass optical filter in the detection channel that transmits only the higher wavelengths. If the T_1 excitation energy is equal to half of the S_1 energy or less, an encounter of a ground state molecule with an excited S_1 molecule can result in singlet fission, as e.g. in:

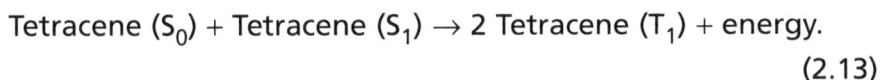

$$\text{Tetracene } (S_0) + \text{Tetracene } (S_1) \rightarrow 2 \text{ Tetracene } (T_1) + \text{energy.}$$

$$(2.13)$$

The two triplet states couple to an overall spin singlet so that the process involves a mutual exchange of two electrons rather than a spin flip. It is therefore fast and distinct from ISC. If there is a small S_1–T_1 energy difference, this allows thermally activated reverse ISC, which then leads to delayed fluorescence[13] and thus to two different lifetimes in the fluorescence spectrum.

2.7 The Franck–Condon Principle

The process of photon absorption occurs on a time scale much faster than nuclear motion. The excited state is from a molecular structure that has identical bond lengths and bond angles as the ground state. Subsequently, due to the rearranged electron distribution with more antibonding character in the excited state, the nuclei will adapt their positions. This fact commonly reflects that some of the bonds are weakened and therefore become longer by an amount q_{01} for the (mass-weighted) nuclear coordinate displayed in Figure 2.10. We have already briefly addressed the vibrational substructure in the absorption and emission spectra of Figure 2.9. In order to understand the intensity distribution of this substructure in which the vibrational quantum number n changes simultaneously with the electronic state, we have to amend the wave function in the formulation for the electronic transition moment μ_{fi} between final state f and initial state i (Eq. (2.8)). We assume that the electron and the nuclear motions are independent since they occur at drastically different time scales because of their drastically different masses. In this case, the vibrational and the electronic energies are additive. We can write the complete wave function as a product of the electronic wave function $\psi_e(r)$, the nuclear wave function ψ_n (R), and for completeness, the spin wave function ψ_s. Also, the transition dipole moment for the electric dipole transition is given as the sum of an electronic contribution μ_e and a nuclear contribution μ_n (neglecting here the contribution for the magnetic transition dipole moment that acts on the spin coordinates and is important only in magnetic resonance transitions).

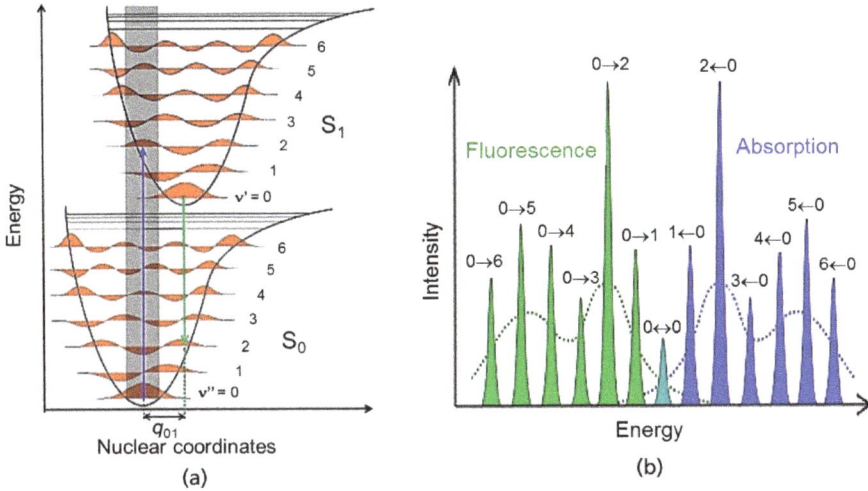

Figure 2.10: Schematic explanation of the Franck–Condon principle: (a) Energy diagram with vibrational sublevels and their corresponding vibrational potentials and wave functions for the S_0 and S_1 electronic states for the case where the equilibrium bond length in the excited state has increased by q_{01}. The shaded area indicates the vibrational wave function overlap of the S_0 vibrational ground state with those of the S_1 excited state. (b) Intensity distribution of the vibrational fine structure in the absorption and fluorescence spectrum as derived from the overlap of vibrational wave functions in (a). Arrows pointing to the right represent the spontaneous emission, those to the left (stimulated) absorption. Note the mirror symmetry between absorption and fluorescence if plotted against energy (not against wavelength λ) (adapted from Ref. [14]).

$$\mu_{fi} = \int \psi_{fe}^* \psi_{ne}^* \psi_{se}^* (\mu_e + \mu_n) \psi_{ie} \psi_{in} \psi_{is} d\tau_e d\tau_n d\tau_s$$

$$= \underbrace{\int \psi_{fe}^* \mu_e \psi_{ie} d\tau_e} \underbrace{\int \psi_{fn}^* \psi_{in} d\tau_n} \underbrace{\int \psi_{fs}^* \psi_{is} d\tau_s} + \underbrace{\int \psi_{fe}^* \psi_{ie} d\tau_e} \underbrace{\int \psi_{fn}^* \mu_n \psi_{in} d\tau_n} \underbrace{\int \psi_{fs}^* \psi_{is} d\tau_s}$$

| orbital selection rule | Franck–Condon factor | spin normalisat. 1 | orbit orthogonality 0 | nuclear transition moment | spin normalisat. 1 |

$$= \int \psi_{fe}^* \mu_e \psi_{ie} d\tau_e \int \psi_{fn}^* \psi_{in} d\tau_n. \tag{2.14}$$

From Eq. (2.14) it can be seen that the integral over the spin functions is the normalisation integral and therefore equals unity because the initial and the final state are both singlet states. Furthermore, the integral over the electron coordinates in the second term of the sum is zero since the initial and the final states are orthogonal. What is left is a product of only two integrals, the one of Eq. (2.8), the orbital selection rule, multiplied with the overlap integral over the nuclear (vibrational) wave functions of the initial and the final states, which is called the Franck–Condon factor.

It is now important to realise that the $3N-6$ vibrational degrees of freedom of an N-atomic molecule are identical for the electronic ground state and the excited state. For example, the electron density in a C–H bond does not change significantly since its electrons are essentially not involved and the vibrational frequency remains the same after excitation. Initially, the vibrational degrees of freedom are mostly in the ground state ($v'' = 0$), and its wave function overlaps only with the ground state vibrational function ($v'' = 0$) of the final electronic state. The Franck–Condon factor of the $0 \leftrightarrow 0$ transition is then unity, and for all other values of v'' it is zero. These vibrational degrees of freedom therefore do not contribute to a vibrational shift of the electronic transition.

There is often only one or very few vibrational degrees of freedom for which this is not the case. They are the ones which involve stretching of C–C bonds of the delocalised π system of the chromophore, mostly the ones parallel to the electronic transition moment since they undergo the most noticeable change in bonding character upon excitation. This changes not only the frequency of the vibration but also the bond length, which results in a displacement of the potential energy minimum of the vibration by an amount q_{01} (see Figure 2.10). These vibrations of the excited state are no longer orthogonal to those of the ground state. It affects the Franck–Condon integrals and allows simultaneous excitation of vibrations $i \leftarrow 0$ of v''. It leads to a fine-structure of electronic bands containing shifts due to only a single or a few vibrations.

The Franck–Condon factor represents the overlap of the vibrational wave functions between the initial and the final states. This is indicated by the shaded area in Figure 2.10 for absorption from the vibrational ground state ($v'' = 0$), which is the only one that is significantly populated at room temperature. The maximum overlap is obtained for the wave function of $v'' = 2$, the $2{\leftarrow}0$ sub-transition is therefore the most intense (blue arrow). Fluorescence occurs after vibrational relaxation from the ground state ($v' = 0$), and the overlap is maximal with $v' = 2$ (green arrow, not shaded), so the most intense sub-transition is $0{\rightarrow}2$. Similar arguments hold for the other sub-transitions, which explains the often observed mirror symmetry between absorption and emission. Indicated with broken lines in Figure 2.10(b) is the case when the vibrational structure is incompletely resolved.

2.8 Solvatochromic Effects

A solvent stabilises a solute dye molecule in the ground state as well as in the excited state, but often not to the same extent. Therefore, excitation energy changes cause the band maximum to shift. Stabilisation due to dispersion interactions is nearly independent of the excitation state, since it relates mostly to charge polarisation of neighbouring solute and solvent atoms. However, excitation changes the overall charge distribution in the solute molecule and therefore its polarity. Thereby, due to the fast nature of the absorptive transition — as described by the Franck–Condon effect — the relevant initial solvent orientation around the S_1 state is the same as that of the ground state, and not in equilibrium. If S_1 is more polar and therefore more strongly solvated than S_0, this will lead to a bathochromic shift, i.e. a shift to longer wavelengths (also called a redshift). The opposite case will be a hypsochromic shift to shorter wavelengths, i.e. a blueshift (Figure 2.11(a)).

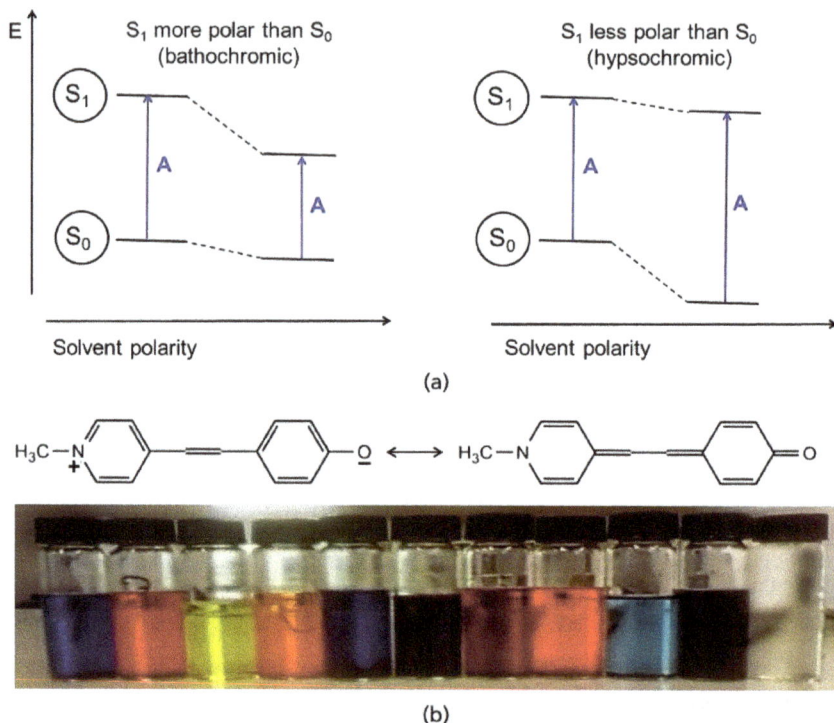

Figure 2.11: (a) Positive/bathochromic (left) and negative/hypsochromic (right) solvatochromism leading to increasing shifts of absorption bands with solvent polarity. (b) Solutions of Brooker's merocyanine dye (MOED) in several solvents, demonstrating large hypsochromic shifts due to preferential solvation of the S_0 ground state (adapted from Ref. [17]).

The molecule 1-methyl-4-[(oxocyclohexadienylidene)ethylidene]-1,4-dihydropyridine, also called Brooker's merocyanine (conventionally abbreviated as MOED), exists in polar solvents a zwitterionic form but in a non-polar solvent in its neutral form.[15] It exhibits spectacular hypsochromic shifts, being yellow (i.e. absorbing blue at 435–480 nm) in water but purple–blue in the less polar acetone (corresponding to absorption at 560–595 nm; Figure 2.11(b)).

Reichardt and Dimrodt defined a quantitative scale for the characterisation of solvatochromic effects. It is based on the pyridine-N-phenoxy-betaine dye, which is a zwitter ion, exhibiting a huge hypsochromic shift from 810 nm in diphenyl ether to 453 nm in water.[16]

Based on the description in Figure 2.11(a) the 0-0 fluorescence band should be expected to occur at the same wavelength as the 0-0 absorption band. In the experiment, however, the florescence bands are normally found to be displaced to longer wavelengths. The origin of this shift, called the Stokes shift, traces back to the time scale of fluorescence (typically several nanoseconds) which allows the S_1 Franck–Condon state to relax to a new equilibrium and thereby lower its energy. Furthermore, the emission process itself is fast and ends in a new non-equilibrium Franck–Condon state that is energetically higher than the equilibrium ground state. Both effects contribute to the Stokes shift (Figure 2.12). Moreover, since the maximum of the absorption band may correspond to a different vibrational sub-state than the maximum of the fluorescence band, this may lead to a further contribution to the Stokes shift. The total effect typically amounts to 20–50 nm for organic dyes in solution. Interestingly, a Stokes shift that depends on the size of nanocrystals was reported for the perovskite $CsPbBr_3$

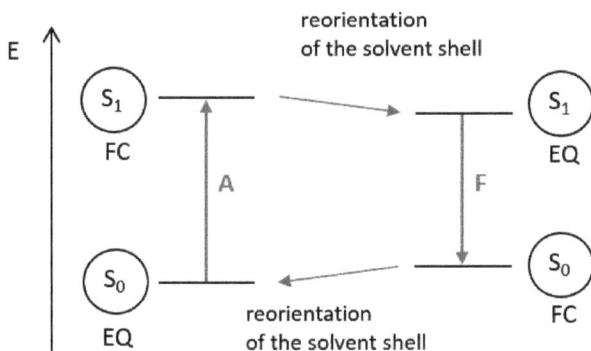

Figure 2.12: Energetic stabilisation of the final state of absorption (S_1) and of fluorescence (S_0) from the initial Franck–Condon (FC) to the equilibrium (EQ) state by reorientation of the solvent shell. Both solvent relaxation processes contribute to the Stokes shift, i.e. the redshift of the maximum of the absorption band to the maximum of the emission band.

and ascribed to a quantum confinement effect of the hole after excitation.[18]

2.9 Determination of Quantum Yields and Lifetimes

As demonstrated in Figure 2.9, an excited state S_1 may deactivate via different decay channels, for example *via* non-radiative processes with rate constant k_{nr}, radiatively *via* fluorescence (rate constant k_F), and via ISC to a triplet state (rate constant k_{ISC}):

$$S_1 \xrightarrow{k_{nr}} S_0, \qquad \text{non-radiative,} \qquad (2.15a)$$

$$S_1 \xrightarrow{k_F} S_0, \qquad \text{fluorescence,} \qquad (2.15b)$$

$$S_1 \xrightarrow{k_{ISC}} T_1, \qquad \text{intersystem crossing.} \qquad (2.15c)$$

These are all first-order kinetic processes which depend only on the concentration $[S_1]$ and the rate constants. The system of S_1 states may be compared with a bucket of water that has three holes of different size. The total loss of water equals the sum of fluxes of water through the holes, $k_{tot} = k_{nr} + k_F + k_{ISC}$. The quantum yields Q_i represent the fractions of excited states which deactivate via a given channel (the fraction of water that is lost though one of the holes):

$$Q_{nr} = \frac{k_{nr}}{k_{nr} + k_F + k_{ISC}}, \qquad (2.16a)$$

$$Q_F = \frac{k_F}{k_{nr} + k_F + k_{ISC}}, \qquad (2.16b)$$

$$Q_{ISC} = \frac{k_{ISC}}{k_{nr} + k_F + k_{ISC}}. \qquad (2.16c)$$

A formula for the experimental determination of fluorescence quantum yields is given in Section 5.3.3 (Eq. (5.1)). The sum of all quantum yields is unity. For their determination we have to count the photons. The number of photons which are absorbed can be obtained by calibration of the photon flux of the excitation source and using the Lambert–Beer Law. An accurate absolute determination is quite tricky, but fortunately we need only relative values. The quantum yield of fluorescence is the ratio of emitted photons in a fluorescence band, relative to the number of absorbed photons. Unless the emission is isotropic the fluorescence quanta have to be integrated over the spectral range of the band and over all spatial directions of emission using an Ulbricht sphere. Assuming that the detectors have the same efficiencies over the different spectral range of absorption and fluorescence this may provide useful estimates. A convenient and perhaps better way is a comparison of the integrated fluorescence with that of tabulated fluorescence standards such as Rhodamine and correction with the squared index of refraction of the medium (see Section 3.3.3).[19] Phosphorescence can be measured the same way if it is sufficiently separated in the spectrum from fluorescence, otherwise one has to resort to pulsed excitation and gated detection (Section 3.5.2). Non-radiative deactivation is a dark channel that can only be assessed experimentally *via* the difference between absorbed and emitted (sum of fluorescence and phosphorescence) photons. The number of phosphorescence quanta represents a lower limit on the ISC, since triplet states may also partly deactivate non-radiatively.

The above processes (2.15a)–(2.15c) are all of first order. Their rate constants have the unit s^{-1}, and the inverse rate constant k_{tot}^{-1} is equal to the lifetime of the exponential decay curve, τ_{tot}:

$$\tau_{tot} = k_{tot}^{-1}, \ \tau_F = k_F^{-1}, \ \tau_{nr} = k_{nr}^{-1}, \ \tau_{ISC} = k_{ISC}^{-1}. \tag{2.17}$$

It can be measured in time-dependent experiments with pulsed excitation at sub-nanosecond resolution, for example using Time Correlated Single Photon Counting (TCSPC, see Section 3.5) experiments. Note that the rates, which for first-order processes are identical to the rate constants, are additive, but the lifetimes are not. The quantum yields can also be expressed in terms of the lifetimes:

$$Q_{nr} = \frac{\tau_{tot}}{\tau_{nr}}, \qquad (2.18a)$$

$$Q_F = \frac{\tau_{tot}}{\tau_F}, \qquad (2.18b)$$

$$Q_{ISC} = \frac{\tau_{tot}}{\tau_{ISC}}. \qquad (2.18c)$$

Section 3.7 will treat the analysis of decay kinetics when competing second-order processes like excited state quenching are involved.

2.10 Defects in Diamond: A Model Solid

Before we proceed to more complex solids like luminescing semiconductors and metals exhibiting plasmon resonances, we discuss diamond as an instructive, conceptually simpler solid that bridges the behaviour between a common molecule and an extended solid. A diamond crystallite can be regarded as a single molecule with all atoms covalently bound, four in tetra-hedral symmetry about a given central carbon atom. A more useful picture is that of a solid consisting of an extended 3-dimensional array of identical face-centred cubic elementary cells with 8 atoms per unit cell. From an optical point of view, intrinsic diamond is entirely featureless and transparent above an absorption edge in deep UV at 225 nm, apart from moder-ate absorption in the infrared between 2.6 and 6.2 μm (266 and 6200 nm).[20] The short-wavelength absorption represents elec-tron excitation from the narrow valence band, perhaps more appropriately called localised valence bond orbitals, across the insulator band gap of 5.5 eV into the delocalised conduction band.

Thus, from the viewpoint of UV–Vis spectroscopy, intrinsic diamond is not interesting. However, in reality, almost no gem-sized natural diamond is absolutely perfect and colourless. There are a variety of common defects in natural as well as synthetic diamonds. Defects generally lead to localised states which are energetically located in the band gap. Donor states arising from defects with excess electrons, like nitrogen that has five valence electrons instead of the four of carbon, lead to states relatively little below the lower edge of the conduction band (Figure 2.13(a)). Acceptor states with less

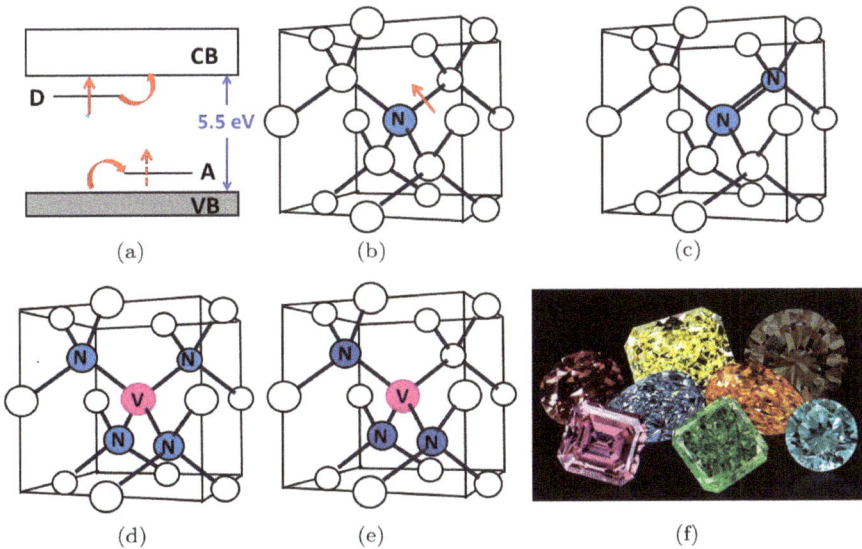

Figure 2.13: (a) Schematic view of donor (D) and acceptor (A) defect states in the gap between valence band (VB) and conduction band (CB) of diamond. (b) Substitutional nitrogen defect (C centre), (c) dinitrogen defect (A centre), (d) vacancy surrounded by four nitrogen atoms (B centre), (e) three nitrogen atoms neighbouring to a vacancy, and (f) selection of natural diamond colours ((b–e) redrawn from Ref. [26], (f) reprinted from Ref. [27]).

than four valence electrons, like boron, are located above the energy of the HOMO orbital (that is equivalent to a narrow valence band). These defect states are localised. By thermal excitation, a donor state can transfer an electron to the conduction band, and an empty acceptor state can get populated by an electron from the valence band. Defect doping by donors and acceptors turns the insulating intrinsic diamond into *n*-type or *p*-type semiconductors, respectively. Simultaneously, doped diamonds adopt typical colours, i.e. they can be excited by absorption of photons in the visible, and they exhibit luminescence.

Colourless diamonds can be grown under controlled conditions, but they are relatively rare in nature. Defects as a result of lattice irregularities or substitutional or interstitial extrinsic single atoms or groups of atoms and vacancies are common. They can be electrically neutral or charged, and they are observable by a variety of spectroscopic methods like EPR (provided that unpaired electron spins are involved), infrared and UV–Vis spectroscopy, in particular photoluminescence.[21,22]

Many of the defects are localised perturbations in a crystal. Their symmetry is therefore described by point groups, just as that of molecules. Since diamond itself has a cubic structure, the various defects possess high symmetries as well.[23] Again in analogy to molecules this has its consequences for the selection rules and the transition moments of the absorptions and emissions, which is of great help for an understanding of the defect structures.

The most common extrinsic defect is nitrogen, and it can comprise up to 1% of the diamond mass.[24] N can be present as isolated atoms surrounded by carbon, or in groups, often combined with a vacancy. A selection of frequently found defects is displayed in Figures 2.13(b)–2.13(e) and a number of attractively coloured natural gem stones in Figure 2.13(f). The simplest one is represented by an isolated N atom in substitutional position (i.e. in the position of a C atom that it replaces, Figure 2.13(b)). This defect is called a C centre, and imposed by its environment it is obviously of tetrahedral symmetry. Since N contributes five valence electrons, one more than C, there is an extra electron that makes the defect a donor that is active in EPR spectroscopy. At a photon energy above 2.2 eV this electron can be excited to the conduction band. One N atom per 100,000 C atoms is sufficient to produce a characteristic yellow colour.[25]

The most common defect in natural diamond results from substitution of two neighbouring C atoms by dinitrogen molecules, which results in a diamagnetic defect of 3-fold symmetry about the body diagonal of the cubic elementary cell (A centre, Figure 2.13(c)). It has an absorption threshold of 310 nm (about 4 eV) and is therefore colourless. This defect is EPR-inactive, however, ionisation by UV excitation leads to an EPR active species.

The B centre is a relatively rare but quite remarkable arrangement of 4 N atoms around a vacancy (V, an empty position in a diamond lattice, Figure 2.13(d)). It is of tetrahedral symmetry and not active in the visible part of the spectrum, thus it is colourless. It is difficult to investigate as it occurs mostly together with the much more frequent A, C and N3 centres. Furthermore, different centres may accidentally absorb at similar spectral positions so that the whole spectrum rather than single bands should be used to characterise defects.[26]

The N3 centre is similar to the B centre, but it contains only 3 N atoms around a vacancy. It is therefore paramagnetic and of 3-fold symmetry. It produces a characteristic absorption and luminescence line at 415.5 nm. The defect commonly occurs together with B centres.

Natural diamonds are found in an impressive spectrum of colours at often high intensities (Figure 2.13(f)). It is often the result of a mixture of defects, including also elements other than N, often B, H, Si, P, Ni, and Co. Interestingly, the colour is quite typical for the origin of the gem stones. Thus, the majority of orange diamonds come from Africa, yellow ones from South Africa, green ones from South America and certain

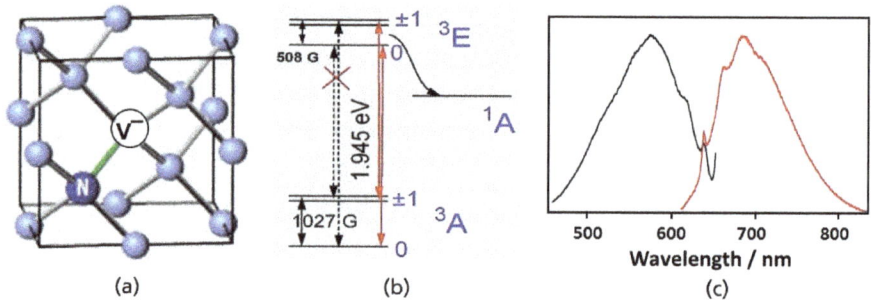

Figure 2.14: (a) Structure of NV⁻ centre, (b) corresponding energy level scheme, and (c) absorption (black) and photoluminescence (red) spectra (reprinted from Ref. [31]).

regions of Africa, blue stones also from the Cullinan mine in South Africa and the Golconda region near Hyderabad in India, pink diamonds mostly from the Argyle mine in Australia (that is also the origin of violet diamonds) but also in Brazil Russia, and southern Africa. Red and purple diamonds are extremely rare.[27]

We now focus on the nitrogen-vacancy (NV) centre where a vacancy is located next to a substitutional nitrogen (Figure 2.14(a)). It is found in the paramagnetic neutral NV^0 and also in the negatively charged NV^- state. The latter is particularly well studied since its quantum properties are suitable for applications such as quantum information processing, single-photon sources and optical magnetometry.[28] In NV^-, the valence electrons from the three carbon dangling bonds around the vacancy combine with the extra electron of the negative charge (and the two nitrogen lone pair electrons) to a ^3A triplet ground state that splits into a doublet and a singlet because of the 3-fold rotational symmetry along the N–V bond (Figure 2.14(b)). Further splitting occurs when a magnetic field is applied. The defect state can be promoted by irradiation at an energy ≥1.945 eV (≤637 nm) to an excited ^3E state that undergoes ISC to a ^1A state which emits photoluminescence. Selection rules dictate that neither the quantum numbers for total spin nor those for its z-component ($m_S = 0, \pm 1$) change during electrical dipole allowed (radiative) transitions. The absorption and luminescence bands are broadened by vibrational progressions (Figure 2.14(c)), just as in organic molecular chromophores where the scaled fluorescence spectrum is often near-symmetric to the absorption spectrum (Figures 2.9 and 2.10). Vibrational progressions due to lattice phonons in defect-doped diamonds were reported in literature and recommended to be used for the identification of the centres.[22]

The absorption cross-section of NV^- was reported to be 0.31 nm^2,[29] which is consistent with the picture developed for molecular chromophores (Eq. (2.7)). The luminescence decay time of the order of 10 ns at room temperature[30] is also similar to typical fluorescence lifetimes (Table 2.2).

Further details about the applications of these defects will be discussed in Section 7.7.

2.11 Band Structure and Optical Processes in Semiconductor Quantum Dots

The above discussion relates to molecular organic chromophores. For semiconductor quantum dots the situation is basically analogous, but several important differences should be noted:

- Organic chromophores often have relatively narrow absorption bands so that different excited S states appear separated, while emission spectra are broadened by a red tail (see Figure 2.6). In contrast, quantum dots have broad absorption spectra, which means that they can be excited by a broad range of wavelengths, but their emission spectra are relatively narrow and symmetric (Figure 2.15).[32]
- Semiconductors have a band gap (the analogue of a HOMO–LUMO gap in molecules) that ranges between less than one and several eV. Large band gap materials are generally found of higher chemical stability because it is more

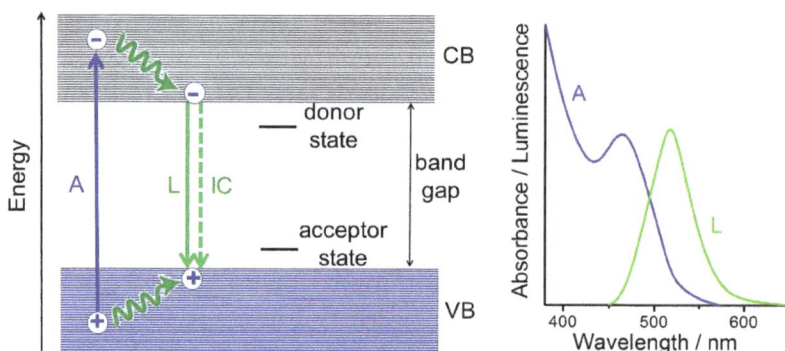

Figure 2.15: Energy diagram of a semiconductor quantum dot. Absorption (A) of a photon occurs from the filled valence band (VB) across the band gap into the empty conduction band (CB). Electron and hole relax towards the band edges (upward relaxation of a hole is equivalent to downward relaxation of an electron to fill the hole) from where they can recombine under emission of luminescence (L), or it dissipates the energy in a non-radiative process by internal conversion (IC). Typical absorption and luminescence spectra are shown on the right. Doping or defect states are localised, not band-like, and normally found inside the band gap that increases as a consequence of the quantum confinement effect as the crystallite size decreases, causing a blueshift of the spectral features, while the energy of the localised defect states is independent of crystallite size.

difficult to remove an electron from the HOMO or to add an electron to the LUMO. At room temperature, thermal energy kT corresponds to 26 meV. Therefore, any material that has a band gap large compared with this value is an insulator in the absence of light. However, doping with electron donor atoms at energies just below the lower band edge of the conduction band can provide electrons in the conduction band by thermal excitation. Alternatively, doping with acceptor atoms just above the upper band edge of the valence band leads to thermal hole formation in the valence band, and therefore to semiconducting behaviour with a positive temperature coefficient of conductivity. A common and abundant defect in oxide materials occurs in the form of HO^- in place of O^{2-}, a remnant of the synthesis process. The necessary extra electron to provide charge neutrality is then found in the conduction band, or it is trapped at a cation, as e.g. in TiO_2 where Ti^{4+} is reduced to Ti^{3+}, a donor state that is located just below the conduction band edge.

- Just like organic chromophores, semiconductor quantum dots are systems with delocalised valence and conduction electrons. These electron states have wave character with nodes at the surface of crystallites. Quantum confinement of the delocalised electrons within the crystallite leads to a strong dependence of the band gap on size, with smaller crystallites having a larger gap, and therefore to a pronounced tunability of the absorption and emission wavelengths by grain size.[33]

- In the dense many-electron bands of quantum dots it is not possible to distinguish states of different spin multiplicity. The rapid interconversion between spin states due to spin–orbit coupling in the heavy atoms prevents the formation of discrete spin states. It is not possible to distinguish between fluorescence and phosphorescence, and the process of light-emitting deactivation is therefore usually called luminescence. The luminescence lifetime is amounts to typically 10–40 ns, an order of magnitude higher than for organic dye molecules.

- An organic chromophore typically consists of a few or several tens of carbon atoms, each contributing with one electron to the planar conjugated π system that forms the chromophore. Inorganic semiconducting quantum dots are 3-dimensional and consist of more atoms or ions, which contribute several atomic orbitals and valence electrons each to a delocalised system that extends over the entire crystallite. A 5 nm CdSe quantum dot with a density of 5.8 g cm^{-3}, for example, consists of 120 CdSe units. On the basis that N atomic orbitals lead to N molecular orbitals which spread roughly over the same energy range, this means that quantum dots provide a significantly higher density of electronic states that are discrete for small quantum dots but form bands that are not resolved experimentally for larger nanoparticles.

- Carbon atoms are light compared with the relatively heavy atoms or ions in a quantum dot. The above example of 120 CdSe units leads to 714 vibrational states (phonons) which are furthermore much more closely spaced than in organic chromophores because of the inverse square root mass dependence of the energy level spacing. These vibrational states are nearly identical for the electronic ground and excited states so that the Franck–Condon factor (Eq. (2.14)) adopts non-zero values only for equal quantum numbers v'' and v'. Therefore, absorption and fluorescence spectra of quantum dots are smooth and do not exhibit any vibrational fine structure. A further origin of the smooth spectra lies in the dependence of the absorption and luminescence wavelengths on crystallite size, which is a quantum confinement effect. Already a very limited distribution of crystallite sizes and shapes in a given sample would serve to wash out any structure in the bands. It is mostly the size distribution that leads to an overlap of absorption and emission spectra (Figure 2.15). This contrasts with organic chromophores where the overlap is due to different solvation of ground and excited state.

- The inorganic systems of quantum dots are resistant to photo-bleaching, which is a clear advantage for their use as long-term luminescent probes. Furthermore, their surface can be grafted for protection against sintering, for solubility, biocompatibility and prevention of luminescence quenching.

2.12 Plasmon Resonances of Metallic Nanoparticles and Surfaces

Nanotechnology is a relatively new science, but gold and silver nanoparticles were fabricated and used already by the

Romans in the 4th Century A.D. as evidenced by the Lycurgus Cup in the British Museum, a glass cage cup that appears of different colour depending on whether it is observed in reflection or in transmission. The colour is due to the gold and silver nanoparticles that were added in colloidal form. Michael Faraday concluded that the variation in colour of finely dispersed gold was determined by the size of the particles that was beyond any power of the microscope, as he reported in his famous Bakerian Lecture on February 5, 1857.[34] A gold sol of red colour that was prepared by Faraday for this lecture is still on display in the British Museum.

The colour of colloidal metal nanoparticles is attributed to plasmon resonances, coherent collective oscillations mainly of conduction electrons in the particle, stimulated by an oscillating electric field (Figure 2.16). The oscillation frequency is determined by four factors: the density of electrons, the effective electron mass, the shape and the size of the charge distribution (and thus normally of the particle).[35] The plasmon generates an evanescent wave that extends outside the particle into the surrounding medium; it is therefore sensitive to the dielectric constant of this medium and to the surface coverage. The frequency can be related to the metal dielectric constant

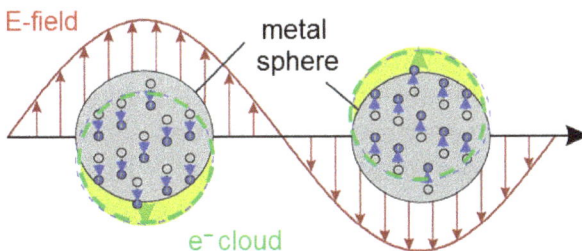

Figure 2.16: A plasmon resonance is a non-propagating collective motion of multiple conduction electrons. In a dipole resonance, the electron cloud is displaced coherently under the influence of an oscillating exciting electric field (redrawn with permission from Ref. [35], © (2003) American Chemical Society).

that can be determined experimentally and is a strong function of wavelength of the bulk metal.[35] For gold and silver, the resonances appear in the visible range of the spectrum.

Apart from the more common dipole plasmon resonances, the electron in larger particles can also oscillate in patterns represented by higher multipoles. In quadrupole modes, half of the electron cloud moves in a direction parallel and half antiparallel to the exciting electric field.[31]

The interaction of light with spherical metal particles with sizes from 10 to 100 nm lead to plasmon resonances by a light scattering process that has been described most successfully by Mie scattering theory.[36] Experimental detection is straightforward by measuring the extinction spectrum (absorption plus scattering) by transmission UV–Vis spectroscopy as a function of wavelength for transparent samples and in reflection mode for non-transparent samples. Just like absorption, scattering reduces the amount of transmitted light. The scattered light can also be detected from single particles using dark field microscopy.

We have to distinguish between light scattering of metallic particles that lead to plasmon resonances and conventional absorption and fluorescence that is dominant in smaller, molecule-like clusters. Plasmon resonances represent changes in the dynamics of conduction electrons at the Fermi level within a continuum band structure. The calculated extinction spectra of oblate silver nanoparticles with different aspect ratios (the ratio of long to short axis) are shown in Figure 2.17, revealing a significant shape dependence of the scattered wavelength.

The intensity of scattered light increases with the 6th power of the nanoparticle diameter. In contrast, the absorption cross-section increases only with the 3rd power of the particle size. Therefore, detection of fluorescence is an alternative to probe smaller particles.[37] Absorption and fluorescence spectra of small gold nanoclusters represent transitions between discrete states (Figure 2.18).[38] Their size-dependence is a consequence of quantum confinement of the delocalised conduction electrons in the cluster, just as in semiconductor quantum dots.

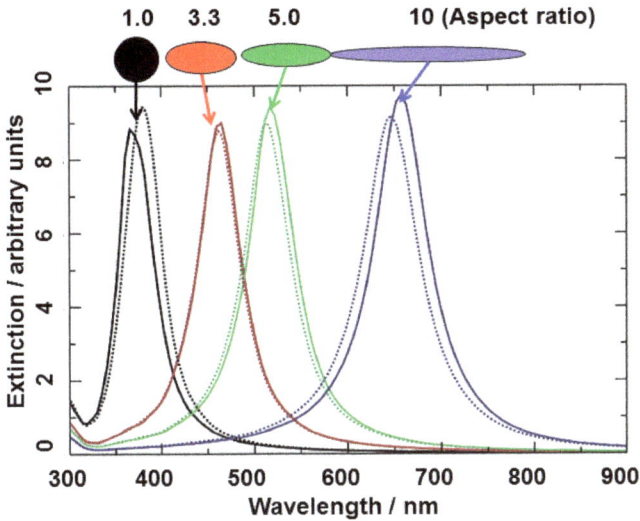

Figure 2.17: Calculated extinction spectra of oblate silver spheroids with different aspect ratios, all with the same equivalent volume corresponding to a sphere of 30 nm radius. Exact solution of the Maxwell equation (solid lines) and approximation (broken lines) (adapted with permission from Ref. [35], © (2003) American Chemical Society).

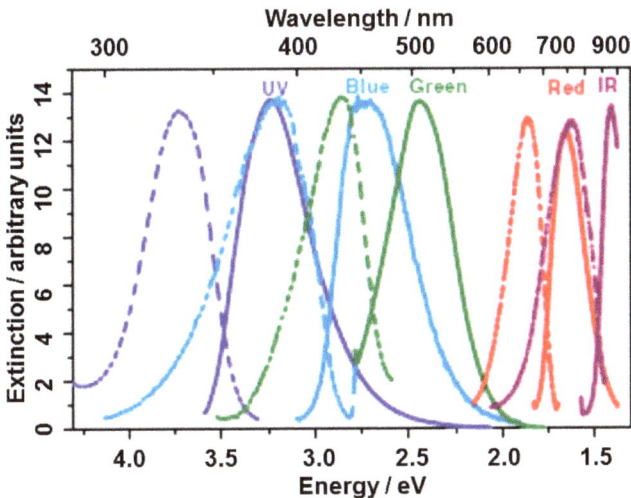

Figure 2.18: Size-tuned extinction (broken lines) and emission (solid lines) spectra of the gold nanoclusters Au_5 (UV), Au_8 (blue), Au_{13} (green), Au_{23} (red), and Au_{31} (near IR). The subscript gives the number of atoms in the cluster. (reprinted with permission from Ref. [38], © (2004) by the American Physical Society).

The ability of metallic nanoparticles to absorb light and form plasmon resonances during the scattering process follows Lambert–Beer's law and can formally be characterised by extinction coefficients just as for electric dipole allowed transitions. For large particles, ε becomes very large. For capped gold nanoparticles with core diameters of 3.76 and 34.5 nm, extinction coefficients $\varepsilon = 3.6 \times 10^6$ and 6.1×10^9 L mol^{-1} cm^{-1}, respectively, were reported.[39] However, since ε scales with the 3rd power of the particle size it is not meaningful to convert ε to a 2-dimensional reaction cross-section σ according to Eq. (2.7).

Plasmon resonances are not restricted to metal nanoparticles but occur also near the surface of the bulk metal or of thin metal films. These are then called surface plasmon resonances (SPRs), and they are extremely sensitive for optical analysis of an analyte that is brought into the vicinity of a metal surface where plasmons are excited by a laser and observed as a function of the scattering angle. The evanescent field extends into the medium above the metal by typically 200 nm, permitting the analysis of changes in small amounts of materials. If the SPR is close to a molecular electronic transition, the evanescent wave also enhances Raman scattering by factors of 10^{11}–10^{12}, leading to a method called surface-enhanced Raman scattering (SERS).[40]

Optically excited smooth metal film surfaces show no or little luminescence, with quantum efficiencies quoted typically as low as 10^{-10}, but more recently, values of 10^{-6} were reported for gold nanoparticles and 10^{-4} for gold nanorods. The origin of this massive enhancement is not yet clear.[41]

2.13 Key Points

- The structure in spectra due to electronic excitation or de-excitation in atoms, molecules or semiconductor quantum dots reflects the occurrence of discrete (non-continuous) energy states that are due to quantum confinement of the electrons in these objects.

- Transitions between electronic states occur spontaneously or are induced by the presence of an oscillating electrical field. The product of the frequency of the absorbed or emitted photon and the Planck constant equals the energy difference between the participating states.
- Absorption in optical spectroscopy is extremely fast and occurs on a time scale in the order of 10^{-15} s. It therefore provides an instant picture of the observed object.
- The absorption (extinction) coefficient in Lambert–Beer's Law is equivalent to an effective absorption cross-section of the absorbing chromophore.
- The final state of an excitation/de-excitation process needs to be empty of half-filled, otherwise the transition cannot occur. Furthermore, particular symmetry of a molecule can suppress certain transitions for which the transition dipole moment, an integral over the two participating wave functions and the exciting electric field vector, becomes zero. Such transitions are called forbidden.
- If in addition to the electronic state also vibrational and/or rotational states change during absorption/emission this leads to a fine structure in the spectra of molecules or solids like diamond, but for semiconductor or metallic nanoparticles no such structure is resolved.
- Deactivation of molecular excited states can be non-radiative by conversion into heat, or radiative by emission of fluorescence from the excited S_1 state, or after ISC by emission of phosphorescence from the T_1 state. Since the phosphorescence quantum yield is often low, the T states are also called dark states. In the presence of heavy atoms or ions, as in semiconductor quantum dots or also in the lead halide perovskites, singlet and triplet states are heavily mixed so that fluorescence and phosphorescence are not distinguishable and are therefore termed luminescence.
- Metallic nanoparticles show SPR, which is a coherent oscillation of multiple electrons near the particle surface. Absorption and emission look similar to that of quantum

dots, but the phenomenon is different and better described by Mie scattering.

General Reading

- J. R. Lakowicz, *Principles of Fluorescence Spectroscopy*, 3rd Ed., Springer, Singapore, 2006.
- M. Smith, S. Nie, Chemical analysis and cellular imaging with quantum dots, *The Analyst*, 2004, 129, 672–677.
- Comin, L. Manna, New materials for tunable plasmonic colloidal nanocrystals, *Chem. Soc. Rev.*, 2014, 43, 3957–3975.
- S. Davydov, *Theory of Molecular Excitons*, Springer, Berlin, 2014, ISBN: 978-1-4899-5171-7.

References

1. http://www.telescope-optics.net/eye_spectral_response.htm.
2. J. Petrusca, *J. Chem. Phys.*, 1961, 34, 1120–1136.
3. J. N. Murrell, J. A. Pople, *Proc. Phys. Soc. A*, 1956, 69, 245–252.
4. J. Li, Ch.-K. Lin, X. Y. Li, Ch. Y. Zhu, S. H. Lin, *Phys. Chem. Chem. Phys.*, 2010, 12, 14967–14976.
5. M. Kasha, H. R. Rawls, M. A. El-Bayoumi, *Pure Appl. Chem., IUPAC*, 1965, 11, 371–399, Online: doi:10.1351/pac196511030371.
6. H. van Amerongen, L. Valkunas, R. van Grondelle, *Photosynthetic Excitons*, World Scientific Publishing, Singapore, 2000.
7. T. P. J. Krüger, *From Disorder to Order: The Functional Flexibility of Single Plant Light-Harvesting Complexes.* Doctoral Thesis, Vrije Universiteit Amsterdam, 2011, ISBN: 978-90-8570-766-0.
8. J. Wrachtrup, T. J. Aartsma, J. Köhler, M. Ketelaars, A. M. van Oijen, M. Matsushita, J. Schmidt, C. Tietz, F. Jelezko, Spectroscopy of individual photosynthetic pigment-protein complexes, Chapter 6. In: *Single Molecule Detection in Solution: Methods and Applications.* C. Zander, J. Enderlein and R. A. Keller (eds.), Wiley-VCH Verlag GmbH & Co. KGaA, Weinheim, FRG, 2002, doi: 10.1002/3527600809.ch6.
9. P. W. Atkins, *Quanta*, Clarendon Press, Oxford, 1974.
10. M. Saba, *Nature*, 2018, 553, 163–164.
11. S. Reinecke, M. A. Baldo, *Sci. Rep.*, 2014, 4, 3797.
12. N. Boens *et al.*, *Anal. Chem.*, 2007, 79, 2137–2149.

13. T. Hait, T. Zhu, D. P. McMahon, T. Van Voorhis, *J. Chem. Theory Comput.* 2016, 12, 3353–3359.
14. https://de.wikipedia.org/wiki/Franck-Condon-Prinzip, downloaded 13.08.2017.
15. J. O. Morley, R. M. Morley, R. Docherty, M. H. Charlton, *J. Am. Chem. Soc.*, 1997, 119, 10192–10202.
16. C. Reichardt, *Chem. Soc. Rev.*, 1992, 21, 147–153.
17. https://en.wikipedia.org/wiki/Brooker%27s_merocyanine, downloaded 12.08.2017.
18. M. C. Brennan, J. E. Herr, T. S. Nguyen-Beck, J. Zinna, S. Draguta, S. Rouvimov, J. Parkhill, M. Kuno, *J. Amer. Chem. Soc.*, 2017, 139, 12201–12208.
19. J. R. Lakowicz, *Principles of Fluorescence Spectroscopy*, 3rd Ed., Springer, New York, 2006.
20. R. P. Mildren, Intrinsic optical properties of diamond. In: *Optical Engineering of Diamond*, R. P. Mildren and J. R. Rabeau (eds.), Wiley-VCH, Weinheim, Germany, 2013.
21. J. Fridrichova, P. Bačík, R. Škoda, P. Antal, *Acta Geologica Slovaca*, 2015, 7, 11–18.
22. H.-C. Lu, B.-M. Cheng, *Anal. Chem.*, 2011, 83, 6539–6544.
23. J. Walker, *Rep. Prog. Phys.*, 1979, 42, 1605–1657.
24. W. Kaiser, W. Bond, *Phys. Rev.*, 1959, 115, 857–863.
25. K. Nassau, *Gems Made by Man*, Geological Institute of America, Santa Monica, California, p. 191.
26. https://en.wikipedia.org/wiki/Crystallographic_defects_in_diamond, downloaded 26.01.2018.
27. http://www.ncdia.com/about_us.php, downloaded 24.03.2018.
28. I. Aharonovich, S. Prawer, *Quantum Information Processing with Diamond, Principles and Applications*, Woodhead Publishing Series in Electronic and Optical Materials, Elsevier, 2014.
29. T.-L. Wee, Y.-K. Tzeng, C.-C. Han, H.-C. Chang, W. Fann, J.-H. Hsu, K.-M. Chen, Y.-C. Yu, *J. Phys. Chem. A*, 2007, 111, 9379–9386.
30. A. T. Collins, M. F. Thomaz, M. I. Jorge, *J. Phys. C*, 1983, 16, 2177–2181.
31. https://en.wikipedia.org/wiki/Nitrogen-vacancy_center, downloaded 27.01.2018.
32. A. P. Alivisatos, W. Gu, C. Larabell, *Annu. Rev. Biomed. Eng.*, 2005, 7, 55–76.
33. E. Roduner, *Chem. Soc. Rev.*, 2006, 35, 583–592.

34. M. Faraday, *Phil. Trans. Royal Soc. London*, 1857, 147, 145–181.
35. K. L. Kelly, E. Coronado, L. L. Zhao, G. C. Schatz, *J. Phys. Chem. B*, 2003, 107, 668–677.
36. U. Kreibig, M. Vollmer, *Optical Properties of Metal Clusters*, Springer Series in Materials Science 25, Springer Berlin, Heidelberg, New York, 1995.
37. C. T. Yuan, W. C. Chou, J. Tang, W. H. Lin, W. H. Chang, J. L. Shen, D. S. Chuu, *Opt. Exp.*, 2009, 17, 16111–16118.
38. J. Zheng, C. Zhang, R. M. Dickson, *Phys. Rev. Lett.*, 2004, 93, 077402.
39. X. Liu, M. Atwater, J. Wang, Q. Huo, *Colloids Surfaces B: Biointerfaces*, 2007, 58, 3–7.
40. K. A. Willets, R. P. Van Duyne, *Ann. Rev. Phys. Chem.*, 2007, 58, 267–297.
41. E. Dulkeith, T. Niedereichholz, T. A. Klar, J. Feldmann, G. von Plessen, D. I. Gittis, K. S. Mayya, F. Caruso, *Phys. Rev. B*, 2004, 70, 205424.

Chapter 3

Aspects of Experimental Setup and Data Analysis

3.1 Introduction

Optical spectroscopy is the experimental technique that offers the widest range of accessible time scales, from steady-state to sub-femtoseconds ($<10^{-15}$ s). Time-resolved spectroscopy allows the investigation of an extremely wide range of processes and materials. For example, one can follow the transient dynamics of electron clouds in photoactive molecules during a reaction, exciton dynamics in semiconductors and organic molecules, and even the motion of electrons around nuclei.

A wide range of commercial instruments are available for steady-state spectroscopy and for a number of time-resolved spectroscopy techniques. Understanding the operating principles behind the commercial instruments is not only useful in deciding on the best equipment for a particular budget but it is also important for identifying possible sources of error, for taking measures against interference by scattering, and for optimising one's results. For example, for samples with a very weak signal the sensitivity of the instrument has to be optimised without compromising the properties of the sample. It is also important to know when and how measured spectra should be corrected. Most published steady-state fluorescence spectra are not correct and cannot be reproduced exactly on a different instrument. Measuring absolute intensities in polarisation and time-resolved techniques is also technically complicated.

Knowledge of one's instrument will help to avoid potential pitfalls and allow for necessary corrections to be made to the data.

In this chapter, the basic units of a steady-state absorption and fluorescence spectrometer are discussed, sources of error in fluorescence spectra are highlighted and methods to correct these spectra are outlined. Diffuse reflectance spectroscopy (DRS) is described as an alternative to absorption spectroscopy when the latter is challenging or impossible to perform. Thereafter, the strengths and pitfalls of polarisation measurements are highlighted, and, finally, various data analysis approaches are introduced.

3.2 Steady-state Absorption Spectroscopy

The central aspect of absorption spectroscopy is illustrated in Figure 2.3. An absorption spectrum is constructed by measuring I_0 and I, the light intensity before and after the sample, respectively, for each wavelength. The transmittance T at each wavelength is then calculated from the ratio $T = I/I_0$, and the absorbance, or optical density, is given by

$$A = -\log T = \log \frac{I_0}{I}. \tag{3.1}$$

Because this method depends critically on the transmitted light it can only be used for optically transparent samples such as solutions and thin films. Absorption spectra of optically opaque or highly scattering samples can be determined from the reflected light. When the absorption properties of a smooth surface are of interest, the intensity of the sample's specular reflectance can replace I in Eq. (3.1) and the absorbance can be calculated in the same way as for transmitted light. It is more common to measure diffusely reflected light, which is not merely reflected off the surface but also from within the sample. To accurately determine the absorption spectrum from diffusely reflected light is complex because of the effect of scattering. DRS will be described in Section 3.4. In this section, the straight-line geometry on which the law of Lambert–Beer is based will be described for light transmission through a sample.

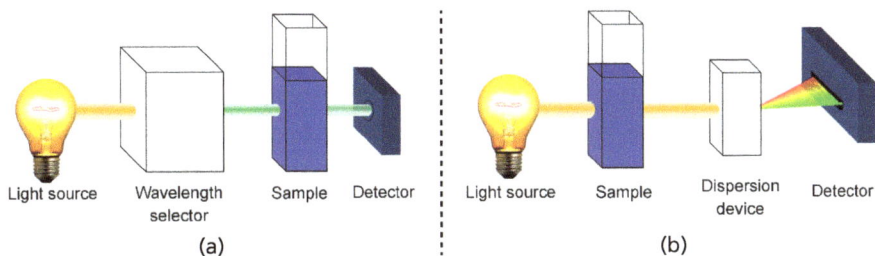

Figure 3.1: Key, indispensable components of a single-beam optical absorption spectrophotometer, using a single-channel detector (a) or a multichannel detector (b).

The instrument used to measure steady-state absorption spectra is known as a spectrophotometer. Two optical layouts can be used. The basic units of these layouts are shown in Figure 3.1. In the first scheme (Figure 3.1(a)), the spectrophotometer scans through a broad wavelength region and measures the absorption for each small wavelength interval. To this end, a broadband light source, tuneable wavelength selector, and broadband, light-sensitive, single-channel detector are used. In the second scheme (Figure 3.1(b)), the whole absorption spectrum is measured at once. For this purpose, a multichannel detector is required. The combination of the dispersion device and detector is known as a spectrometer. Each of these components will be discussed.

3.2.1 *Light source*

The two most commonly used light sources are a xenon lamp and a tungsten halogen lamp. Both emit very broad, continuous and stable spectra. High-pressure xenon lamps can be used for middle- to near-UV, visible, and NIR excitation. A tungsten halogen lamp exhibits a more uniform spectral profile but covers only the visible and NIR. For UV excitation, it should be supplemented with an additional source, usually a deuterium lamp. The often-used quartz casing of the lamps allow transmission above 200 nm. Xenon lamps exhibit more intensity fluctuations than halogen and deuterium lamps, which are a source of flicker noise. This noise can be eliminated to a large degree by modulating the intensity, using a xenon flash lamp or an optical chopper. Intensity modulation at the same time diminishes photodamage of the sample because the sample is not continuously illuminated. A xenon flash lamp offers additional advantages of being brighter in the mid-UV and emitting more stable spectra, though the spectra are more structured than those from continuous xenon lamps. Structured spectra from the light source are in general no concern for absorption spectroscopy but the structure can be imprinted on emission spectra if not adequately corrected for.

Mercury vapour lamps represent a good alternative as excitation source for the middle-UV to visible regions, producing a stronger output intensity than xenon and tungsten lamps. They additionally emit strong line spectra at several wavelengths in the UV and visible regions. Although these lines may saturate the absorption spectrum if not adequately filtered, they represent favourable wavelengths for selective excitation of commonly used fluorophores and can also be used to calibrate the spectrophotometer.[1,2]

3.2.2 *Wavelength selector*

In order to select a single wavelength or narrow section from a broad spectrum, the light beam has to be dispersed into its various colours. A diffraction grating and a dispersion prism are the smallest and most economical devices for this purpose and are therefore typically used in spectrometers. Placing the grating or prism on a rotating table between two slits provides tuneable spectral resolution. These are the main components of a monochromator. A simple layout is given in Figure 3.2. The entrance slit ensures that a single, narrow beam of light enters the monochromator. After the dispersion device (grating or prism) each colour travels in a slightly different direction and

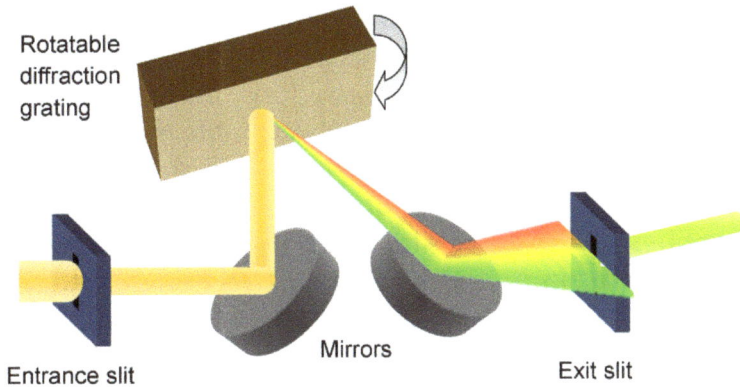

Figure 3.2: Key components of a monochromator.

hits the exit slit at a different horizontal position. By rotating the dispersion device, the spectrum moves horizontally and a different colour can be selected to pass through the exit slit. The slits also serve to reduce stray light, which is light with a different wavelength than the one that passes through the exit slit at a given time.

In the visible region, a prism has a smaller dispersion angle than a grating. A beam of light dispersed by a prism should therefore travel a longer distance before reaching the exit slit than when it is diffracted by a grating and the same wavelength resolution ought to be obtained. Prism monochromators are therefore substantially bigger than grating monochromators. In the past, gratings used to be much less efficient than prisms, and for this reason, old optical monochromators typically contain a prism, while most modern devices use a grating, owing to great advances in grating technology. In the UV region, prisms generally have a greater dispersion angle than gratings due to the increase of dispersion with the frequency of light. Modern monochromators that are designed to operate mostly in the far-UV typically use a prism as the dispersion device.

Optical gratings produce a significant amount of stray light due to the presence of zero-order and higher-order beams as well as surface roughness. Stray light cannot be completely removed from the beam that exits a monochromator. Filters are generally placed at appropriate positions inside the monochromator to block most of the stray light. Double monochromators (two monochromators used in series) reject a greater amount of stray light but at the cost of a lower output intensity. The same applies to a smaller slit width. Use of a concave grating instead of a straight grating can further reduce stray light, because a straight grating should be used together with lenses or concave mirrors, and a higher number of reflecting surfaces give rise to more stray light. The amount of stray light co-propagating with particular wavelengths can be determined by blocking the beam with various band-block filters and comparing the measured spectra with those obtained using unfiltered light. Alternatively, the linearity of the instrument at low sample concentrations can be determined. This is done by calculating the deviation from a linear relationship between the sample concentration and absorbance at selected wavelengths. A perfect linear relationship indicates that the amount of stray light is negligible.

The spectral resolution of a commercial monochromator is tuned by adjusting the exit slit width. A smaller width increases the spectral resolution but lowers the intensity. The user should therefore find a trade-off between spectral resolution and intensity. If the option exists, a higher diffraction order of the grating can be used or the grating can be replaced with one having a larger groove density. Both changes will enhance the spectral resolution but at the cost of decreasing both the intensity and the spectral bandwidth (i.e. the accessible wavelength interval). Alternatively, the spectral resolution can be increased using a monochromator with larger physical dimensions or using multiple monochromators.

An alternative to a monochromator is an acousto-optical tuneable filter. This device uses sound waves to create a grating that diffracts a narrow wavelength band of the incident light. The output wavelength can be tuned by changing the frequency of the incident sound waves. This tuneable filter can operate at a high speed and provide random or sequential wavelength access with high spectral resolution. The drawback is its high cost.

3.2.3 *Sample chamber*

In a steady-state spectroscopy experiment, the sample chamber is the greatest potential source of noise. Awareness of common sources of noise is the first step to avoiding them.

First, a clean working environment is essential for sample purity, longevity of the spectrophotometer, and for limited scattering due to dust. If the sample compartment does not contain windows towards the monochromator and detector, dust can accumulate on the optical components and will not only attenuate the beam but may also cause burns on the optical components when hit by a high-intensity laser beam. Windows in the sample compartment should be regularly checked for dust.

Second, contaminants in the sample should be avoided. This includes small bubbles that may be formed by violent mixing of the sample. Light-sensitive contaminants distort the spectra, while all other contaminants may scatter the light. Reflected and scattered light gives rise to erroneous data because the Lambert–Beer law is based on negligible reflection and scattering. Most of this light does not reach the detector and is therefore interpreted as absorption, thus overestimating the amount of absorption. Furthermore, scattering is strongly wavelength dependent and consequently skews the absorption spectrum.

Third, various considerations have to be taken into account regarding the sample holder. A liquid sample is typically placed in a cuvette. The quality and condition of the cuvette are critical for reliable data. For example, stains will cause scattering and possibly absorption of the incident light. Plastic and ordinary glass cuvettes are suitable for the visible region while quartz cuvettes provide much better transmission in the UV.

The cuvette should be oriented perpendicularly to the incident beam, otherwise the beam is displaced transversely and less light may hit the detector. Care should be taken that the exciting beam does not hit any surface of the liquid, because this will result in scattering.

Fourth, measurement of a reference sample is essential. It would be wrong to calculate the absorbance considering I_0 to represent the light intensity after the monochromator and I the light intensity after the sample in the cuvette. For such a setup the spectral properties of the solution and cuvette are not taken into account. I_0 is rather measured using a reference sample, comprising an identical cuvette filled with the same solvent as for the sample. The spectrum from the reference sample is known as the baseline. Most components of the spectrophotometer, in particular the light source, gratings, and detector, are wavelength dependent. This means that the emission intensity of the light source, transmission efficiency of the monochromator and detection efficiency of the detector are not the same for different wavelengths. Since the wavelength dependence of these components is independent of intensity, the ratio I/I_0 is independent of the wavelength, and so are the absorbance (A) and transmittance (T), because they are calculated using I/I_0. As a result, the spectrophotometer's wavelength dependence is automatically corrected for and no additional corrections are needed. From this discussion it should also be clear why an accurate reference spectrum is as important as an accurate sample spectrum.

Finally, a double-beam configuration is recommendable. In a single-beam spectrometer, the reference and sample spectra are measured consecutively. The drawback of this configuration is that it relies on the stability of both the setup and the sample. Changes between the two measurements such as temporal intensity fluctuations of the light source or altered light conditions in the sample chamber will produce deviations in the calculated absorption spectrum. This becomes particularly important when long measuring times are used for weakly absorbing samples. A double-beam configuration, whereby the reference and sample spectra are measured simultaneously, circumvent this problem.

3.2.4 *Light detector*

For absorbance measurements, a light-sensitive device is required that converts the incident light into a measurable electronic signal. For the configuration in Figure 3.1(a), a single-channel detector, also known as a point detector, is sufficient. Multichannel detectors are discussed in Section 3.2.5.

Photomultiplier tubes (PMTs) and silicon photodiodes are commonly used for UV–Vis spectroscopy. Photodiodes are smaller and cheaper, while PMTs offer a higher sensitivity and less noise and are therefore the choice for high-grade instruments.

When small absorption signals are measured the detector noise may become comparatively high. Noise originating from photon detection and signal conversion is random and cannot be corrected for or changed but the detector's dark current can. The dark current is the current in the absence of incident light and produces a baseline offset for any measured spectrum. It can distort the spectrum of even strong signals and can fluctuate and drift for example due to temperature changes of the detector. It is therefore good practice to measure a dark spectrum before every measurement. A dark spectrum is measured using identical conditions as for the sample (for example acquisition time) but with the light blocked. If a reference sample is measured separately, a dark spectrum should preferably be measured for both the reference sample and the true sample without delay between the measurements to ensure minimal baseline fluctuations. The corrected absorbance is then given by the equation

$$A = \log \frac{I_0 - I_{D_0}}{I - I_D},$$

\hfill (3.2)

where D_0 and D refer to the dark spectra of the reference and true samples, respectively.

3.2.5 *Spectrometer*

To measure the whole absorption spectrum at once (Figure 3.1(b)) the sample needs to be illuminated with a broadband beam and the transmitted light dispersed onto an array of detectors. Such a multichannel detector measures each small wavelength interval of the incident beam simultaneously in a different channel. The main advantage of this configuration is that the full absorption spectrum can be obtained within a short illumination time. By decreasing the measuring time, it becomes less likely that sample or setup changes occur during the measurement and more reliable spectra are obtained. In addition, real-time measurements of kinetic processes on time scales of milliseconds to several seconds can be performed by following the time evolution of the absorption spectrum.

The dispersive element of a spectrometer is generally an optical grating. The most common array detector is a photodiode array. More sensitive but more expensive options are conductive metal oxide semiconductor (CMOS) and charge coupled device (CCD) cameras. Recent technology has pushed the sensitivity and efficiency of CMOS sensors to an almost equal level as those of a CCD sensor, while the former is less expensive and offers considerably faster detection speeds. Finally, spectral information from sub-millimetre sample areas can be obtained by combining the spectrometer with a microscope or simply with an objective lens.

3.3 Steady-state Fluorescence Spectroscopy

It is technically more challenging to obtain a reliable fluorescence spectrum than an absorption spectrum. Good measures have to be taken to exclude artefacts, in particular due to scattering. Consider the following example. An optical density of 0.1 is widely considered to give the optimal sample concentration for fluorescence measurements. This optical density is at the limit of a linear correlation between the excitation and emission intensities, because at higher optical densities the fraction of emitted light that is reabsorbed by the same sample starts to visibly distort the measured spectrum, a phenomenon known as self-absorption. At this optical density, approximately 20% of the light is absorbed. Consider the exceptional situation of a sample with a fluorescence quantum yield of 100% and an instrument capable of collecting 1% of the emitted light, a percentage that is much higher than what most commercial instruments are capable of. This means that the intensity of collected fluorescence is 0.2% that of the exciting light. Now if the fraction of scattered light collected by the instrument is only 0.1% of the ~80% of light transmitted through the sample, it will contribute to 40% of the fluorescence signal!

In this section, the typical experimental configuration for steady-state fluorescence measurements, common sources of noise and general principles to correct the measured spectra will be discussed.

3.3.1 *Experimental setup*

There are two different types of steady-state fluorescence spectra. The first is called an emission spectrum and is recorded using a fixed excitation wavelength, typically corresponding to the sample's absorption maximum, and measuring the full emitted spectrum. The second type is known as an excitation spectrum and is measured by scanning through a selected wavelength region, typically corresponding to the full absorption spectrum, while detecting the fluorescence intensity at a fixed wavelength, usually at the fluorescence maximum.

A fluorescence spectrometer is often referred to as a spectrofluorometer or a fluorimeter. The basic experimental setups for excitation and emission spectroscopy have strong resemblances with the two optical layouts of a spectrophotometer in Figures 3.1(a) and 3.1(b), respectively. The two most important differences are the geometry and the presence of optical filters. Figure 3.3 shows the essential components of a spectrofluorometer.

The right-angle geometry ensures that non-absorbed, non-scattered exciting light is not transferred to the detector. For this geometry, the detection system is arranged such that light from the centre of the cuvette or test tube is detected. Samples can also be measured outside a cuvette or test tube. A thin, transparent sample such as a thin film on a glass substrate can be measured in a similar right-angle geometry by orienting the sample such that its front surface faces the incident beam at a 45° angle and its back surface faces the emission monochromator at the same angle (Figure 3.3, inset). This geometry ensures that reflection off the two surfaces of the sample is directed away from the detector. For a strongly absorbing or an opaque sample, too little light is transmitted through the sample and fluorescence should be measured from the front face. In such a situation, a right-angle geometry may direct too much specular reflection into the detector, and a different angle of incidence

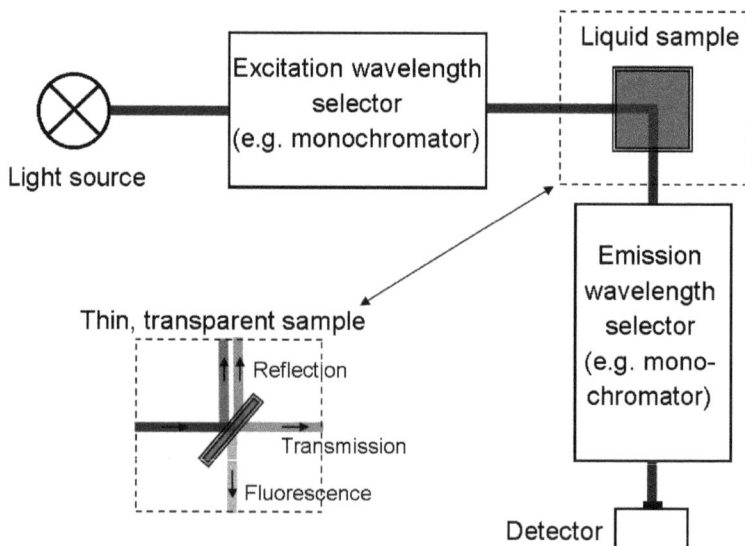

Figure 3.3: Schematic of a spectrofluorometer, viewed from the top.

is often used. The last geometry gives rise to much more scattering into the detector than the first two.

To obtain an excitation spectrum the excitation wavelength should be tuneable. A monochromator is the most economical solution offering high spectral resolution. To select a single colour from the emitted spectrum, a narrow band-pass filter represents the least expensive option. For an emission spectrum the requirements for wavelength selectors are reversed: in the excitation branch, a single band-pass filter can be used to select an excitation wavelength, while in the emission branch the full fluorescence spectrum should be directed into the detector. The fluorescence spectrum can be measured either by scanning the spectrum with a tuneable wavelength selector while detecting the intensity corresponding to each wavelength interval, or by measuring the full spectrum at once using a spectrometer (see Section 3.2.5). One can also choose to scan through both the excitation and emission spectra by measuring the full fluorescence spectrum for each excitation wavelength. This provides a 2-dimensional excitation-emission spectrum.

Most commercial spectrophotometers are equipped with two monochromators — one serving as the excitation wavelength selector and one as the emission wavelength selector. This provides a versatile instrument that enables the measurement of both excitation and emission spectra as well as the selection of any wavelength in the excitation and emission branches. Some companies allow the user to select some of the basic units (the light source, sample compartment, and detector) from a number of standard or even customised options. This is useful to optimise the instrument's performance for particular applications. However, since commercial instruments are usually designed for versatility (i.e. a broad range of applications), the degree of optimisation for a particular application is often limited. The best solution for focussed applications is to build one's own instrument. One such application is single-molecule spectroscopy (see Section 7.1), where the spectrofluorometer should be sensitive enough to accurately measure the fluorescence from single molecules.

The fluorescence signal is often more than 1,000 times weaker than the exciting light. There are various ways to increase the fluorescence intensity.

- First, a suitable light source should be used. For emission spectra, the exciting beam should have a narrow bandwidth, and selecting a narrow band from a spectrally broad exciting light spectrum may give an exciting beam with a too weak intensity. A line source provides considerably higher intensities. Economical options are LEDs and diode lasers. Alternatively, the strong, discrete, narrow emission lines from mercury lamps can be selected by using appropriate band-pass filters. Some commercial systems allow the option of switching between a continuous light source and a line source. Care should be taken that photobleaching does not occur when increasing the exciting light intensity.
- Second, the monochromators can be replaced with appropriate band-pass filters, especially when emission spectral measurements are the main application. The amount of light transmitted through a filter can easily be at least 10 times higher than through a monochromator. Emission spectral bands are often significantly broader than absorption bands. Spectral resolution is then of lesser importance and the emission monochromator can be replaced with a continuously variable band-pass filter, whereby the transmission wavelength band is adjusted by sliding the filter. A courser resolution can be obtained using a collection of band-pass filters mounted on a rotating wheel. Some commercial spectrofluorometers allow the option of bypassing the emission monochromator so that band-pass filters can be used manually. Alternatively, the gratings of the monochromator can be set such that the zeroth diffraction order is used, in which case the full spectrum is transmitted through the monochromator.

3.3.2 *Sources of error*

Interference by light scattering is the most common source of error in fluorescence measurements. For most samples the right-angle geometry of the experimental setup eliminates a sufficient amount of scattering from the detection channel. However, in many situations this is inadequate, for example, for weakly fluorescing or very dilute samples, strongly scattering samples, or strongly absorbing samples where comparatively high levels of scattered light enter the detection channel. Biological samples are often turbid due to their large concentration of macromolecules, lipids, membranes, polymers, cells, or tissue, which contribute to a significant amount of scattering.

For liquid samples exhibiting a weak fluorescence signal, Raman scattering may be detected in addition to Rayleigh scattering. Since the Raman peak is always shifted from the incident light by a fixed wavenumber, it can be recognised by a shifting band in the fluorescence spectrum when using a different excitation wavelength. Rayleigh scattering can be identified as increased absorption when using a shorter excitation wavelength, because the amount of scattering scales with λ^{-4}. Some commercial spectrophotometers have the option of moving the sample very close to the detector in order to reduce the detection of Rayleigh scattering. Such a possibility is particularly valuable for UV measurements.

The use of optical and spatial filters is important for further elimination of scattered light. It is good practice to always use a fluorescence filter after the sample. These filters block all light below a certain wavelength, while light with a longer wavelength is transmitted. Due to the Stokes shift, all exciting light can be blocked, while the full fluorescence spectrum is detected. However, when Raman scatter is present and the Stokes shift of the sample is small, the best filter may still cut out the blue tail of the fluorescence spectrum. It may also be necessary to use a band-pass filter to spectrally clean the exciting beam when using a single excitation wavelength.

A monochromator acts as a spatial light filter by transmitting only the part of the spectrum that is selected to pass through the exit slit. A similar effect can be achieved using an aperture, also known as an iris, which transmits only the light directed through it and blocks all light travelling in other directions. Better selectivity can be obtained by using a small aperture, known as a pinhole, placed between two lenses. The first lens focusses the light through the pinhole and the second one collimates the light or focusses the beam onto the detector. A pinhole size slightly smaller than the focal beam diameter provides the best signal-to-noise ratio.

Other potential sources of error are stray light, the inner filter effect, and fluorescent impurities in the sample. Stray light has been discussed in Section 3.2.2. The inner filter effect refers to the absorption of exciting and/or fluorescence light outside the central region in the sample cell, considering that only fluorescence originating from this central region is detected. This effect is limited by using an adequately low sample concentration, corresponding to an optical density of typically <0.1. Light-sensitive contaminants also contribute to the inner filter effect by absorbing some of the exciting light. Fluorescent impurities normally give rise to a modified fluorescence spectrum when changing the excitation wavelength, because they typically have a different wavelength dependence of excitation than the fluorophores of interest. However, this would not be observed for impurities with excessively broad fluorescence signals.

The contribution of all above-mentioned sources of error can be determined using a reference sample, which is identical to the real sample but does not contain the fluorophores of interest. After eliminating all sources of noise as sufficiently as possible, the reference sample will give a background spectrum that should be subtracted from the real sample's fluorescence spectrum. The solvent used for biological samples is often not a sufficient reference sample because of scatterers and fluorescent components frequently bound to the fluorophores. For example, for solubilised protein-bound fluorophores, the proteins are strong scattering materials and the aromatic aminoacids of the proteins fluoresce in the UV. For such samples, the best practice is to use a non-fluorescent scattering material as a reference sample. The scattering of proteins can be mimicked using colloidal silica or solubilised coffee creamer,[3] while the effect of scattering from tissues can be mimicked by a lipid emulsion.

3.3.3 *Correction of fluorescence spectra*

Absorption spectra are corrected by dividing the sample's measured spectrum with the baseline. This provides a simple method to obtain absolute intensities. In contrast, it takes more effort and precision to correct a fluorescence spectrum, and unless all intensity losses after the sample are carefully accounted for, the fluorescence spectrum gives only relative intensity values. The intensity of a fluorescence spectrum is therefore typically given in arbitrary units.

When only the relative intensities of an emission spectrum are of interest it is often considered unnecessary to correct the spectrum. It is true that the basic units of spectrophotometers from different manufacturers have a wavelength response that is quite similar. However, it is imperative to correct emission spectra when quantitative data is required, for example when fluorescence quantum yields are to be determined and when the absolute response of a reaction due to variations in experimental conditions has to be established. Excitation spectra may be disproportionately distorted when not corrected. The excitation spectrum of a highly diluted sample in which no excitation quenching or energy transfer between different molecules takes place should be identical to its absorption spectrum. As the sample concentration increases, the relative amplitude of the peaks in the excitation spectrum change (Figure 3.4).

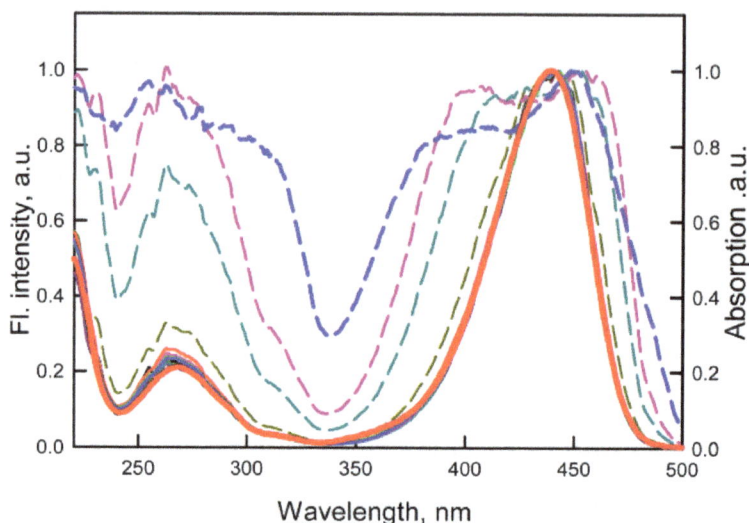

Figure 3.4: Fluorescence excitation spectra of ATTO-425 solutions at various concentrations, before (dashed lines) and after (solid lines) correction for the inner filter effect. The higher the sample concentration, the larger its deviation is from the absorption spectrum (red) (reprinted from Ref. [5], open access, Creative Commons).

A fluorescence spectrum is corrected by dividing the measured spectrum with a correction spectrum, which expresses the relative intensity at each wavelength due to the wavelength dependence of the spectrophotometer. The two strongest sources of error are the light source and the monochromator. The emission spectrum of the light source is not flat, especially in the UV region and for sources like a xenon flash lamp that show a lot of sharp spectral structures. The efficiency of a diffraction grating can also vary strongly for different wavelengths.

Correction spectra are often provided by manufacturers for their spectrophotometers but in some cases this is not sufficient. First, the correction spectrum may only represent the relative wavelength dependence of the detected intensity and not its absolute dependence. The absolute response of the instrument should take into account all intensity losses throughout the instrument. The greatest intensity loss occurs during collection of the emitted light, which may vary for different samples. Second, the spectrophotometer degrades due to aging, affecting its optical performance, especially in the UV. Third, when building one's own spectrofluorometer the new instrument's correction spectrum should be carefully determined.

A correction spectrum can be obtained using a reference sample that has a near-unit fluorescence quantum yield over a broad wavelength range. Such a sample is often called a quantum counter, because it emits essentially the same number of photons that are absorbed, so that its measured spectrum provides a good estimation of the absolute wavelength dependence of the spectrofluorometer. The most reliable approach is to compare the measured spectrum of any standard reference sample with that of a corrected spectrum of the sample found in literature. A list of suitable, standard fluorophores with known absolute emission spectra can be found in several textbooks, for example.[4] High-end commercial spectrofluorometers typically include a quantum counter in a separate compartment so that the measured emission spectrum can be corrected in real time. Such a simultaneous measurement of the quantum counter also accounts for intensity variations of the light source. One disadvantage is that such a built-in quantum counter has to be used for all samples, while its wavelength response is flat over a limited range, which may not correspond with the emission wavelength region of the sample of interest.

Another important correction to perform periodically is calibration. Array detectors and mechanically steered dispersive instruments such as those in a monochromator have to be calibrated. It is advisable to check the emission spectrum of the light source every now and then for signs of degradation.

Finally, for turbid solutions and solid samples, scattered light may still be detected even when using optical and spatial filters. In such a case, the fluorescence data can be corrected using an empirically determined subtraction algorithm.

3.4 Diffuse Reflection Spectroscopy (DRS)

The absorption spectra of samples exhibiting a high degree of scattering are often distorted. This is the case for colloids and other turbid samples, especially when the wavelength is comparable to the mean particle size. Since absorption spectra are commonly determined from the percentage of transmitted light, the spectra from strongly absorbing samples may be too weak and noisy to be accurately resolved. Furthermore, no light is transmitted through opaque samples. For all these samples, the absorption spectrum can be obtained from the reflected light using a technique known as DRS.

Reflectance can be measured across a broad wavelength range, from mid-UV to mid-IR. DRS measurements can therefore be performed using an absorption spectrophotometer, such as diagrammed in Figure 3.1(a), equipped with a diffuse reflectance module in the sample compartment. Two types of modules are commonly used.

The first type is an integrating Ulbricht sphere. The sample is placed inside the highly reflecting sphere, the exciting beam is directed through the entrance port, and the detector is aligned with the exit port.[6] The larger the sphere the more diffusely reflected light from the sample is detected and higher-quality spectra can be obtained. When the exciting light is incident perpendicularly to the sample surface, specular reflected light will exit the sphere through the entrance port and only diffuse reflected light is measured. Using a sphere with a different angle of incidence allows measurements of both specular and diffuse reflected light.

The second type of module makes use of two ellipsoidal mirrors: one for focussing the exciting beam onto the sample and the second for collecting the diffusely scattered light. This method typically collects about 20% of the back-reflected light.[7] A similar idea can be used to measure samples *in situ*, for example biological samples in their native environment. In this case, two optical fibres are used — one for excitation and one for collection of the diffusely reflected light.

The aim of DRS is to reconstruct the absorption spectrum of the sample. The key difference between DRS and transmission-based absorption spectroscopy is the presence of scattering within the sample. For an optically opaque sample, no light transmission occurs, and light that is not absorbed is scattered out of the sample and can be measured. This scattered light, commonly referred to as diffuse reflected light, is therefore equivalent to transmittance in standard absorption spectroscopy. However, the amount of scattering does not scale linearly with the concentration or thickness of the sample like absorption does according to the Lambert–Beer law. Hence, the diffuse reflectance spectrum deviates from the absorption spectrum, especially when relatively little light is transmitted (Figure 3.5(a)). Kubelka and Munk derived a simple equation to calibrate the reflectance spectrum in order to obtain the main features of the true absorption spectrum. For an opaque sample, the diffuse reflectance is given by R_∞, indicating an

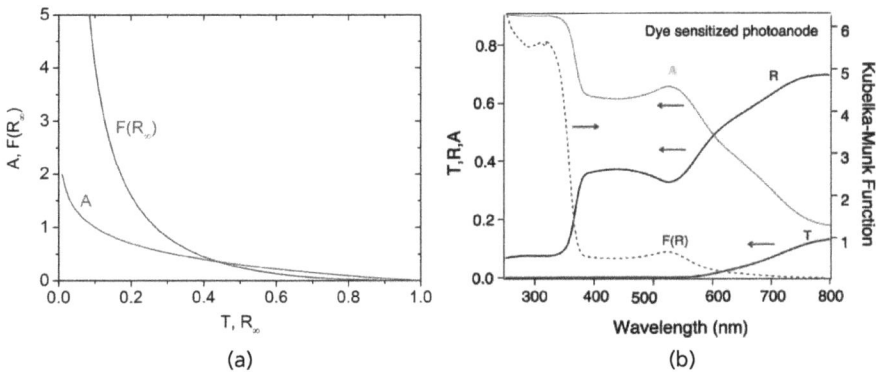

Figure 3.5: (a) Comparison between the absorbance A, and the Kubelka–Munk function $F(R_\infty)$, using Eqs. (3.1) and (3.3). Whereas absorbance is somewhat larger than $F(R_\infty)$ above a transmittance, T, and diffuse reflectance, R, value of 0.44, $F(R_\infty)$ strongly dominates at small values. (b) Parameters of an opaque Ru-complex sensitised TiO_2 double-layer film with an iodide/triiodide based redox electrolyte in 3-methoxypropionitrile solvent. R and T were measured, $F(R)$ was calculated using Eq. (3.3) and A was calculated using the equation $A = 1 - T - R$ (reprinted from Ref. [11]).

infinite optical thickness. The diffuse reflectance spectrum is then transformed or corrected using the following function, known as the Kubelka–Munk function or the remission function[8]:

$$F(R_\infty) = \frac{(1-R_\infty)^2}{2R_\infty} = \frac{K}{S},$$ (3.3)

where K and S are the Kubelka–Munk absorption and scattering coefficients, respectively. R_∞ scales between 0 and 1. For $R_\infty = 0$, no scattering occurs and all light is therefore transmitted or absorbed. For $R_\infty = 1$, no absorption occurs and all light is scattered and diffusely reflected. R_∞ is typically obtained for sample thicknesses of 1–2 mm for solids and up to 5 mm for powders, but the theory still works well for $R_\infty >$ 0.6.[9] The actual value of R_∞ for the sample is obtained by multiplying the sample's measured value with that of a reference sample, i.e. R_∞(sample) = R_∞(measured) × R_∞(reference).[10] The reference sample, also called a white standard, is a sample with a known value of S for all wavelengths. K is directly proportional to the real absorption coefficient of the sample[9] and the reflectance spectrum can therefore be shifted vertically to match the transmission spectrum. Figure 3.5(b) compares the Kubelka–Munk function and absorption spectrum of a dye-sensitised TiO_2 solar cell.[11] In this example, the two spectra show qualitatively the same absorption features but the amplitudes of these features are markedly different because wavelength correction was not performed for the Kubelka–Munk function.

The absolute value of K for each wavelength can be obtained from a series of measurements whereby the sample is mixed with variable amounts of the reference sample, assuming that the latter has the same scattering coefficient as that of the sample of interest. The transmission spectrum of the sample can be measured in a similar way, provided that the reference sample is optically transparent and non-absorbing and a sufficient amount of it is used to dilute the real sample. This will give rise to a measurable amount of transmission while the amount of scattering is assumed to be unaltered. When K is known, Eq. (3.3) can be used to calculate S. Alternatively, the absolute value of S can be determined from the slope of the reflectance vs. sample thickness, using a series of measurements with different sample thicknesses.

Particle size is the factor that most strongly distorts reflectance spectra and hence determination of the absorption spectra. Eq. (3.3) is based on the assumption that K and S are independent of wavelength. This assumption is only valid when the sample's particles are large compared to the wavelength of the incident light. However, when the particles are too large, specular reflection increases, the amount of absorption may become comparatively large, and the particle size heterogeneity is often larger. All these effects lead to a decrease in the amplitudes of the spectral bands, and the bands become more difficult to resolve accurately. Samples are therefore usually ground to particle sizes of 0.1–1 μm.[12] Due to the wavelength-dependent scattering (i.e. increased scattering at shorter wavelengths) of such small particles, the measured spectra should be corrected. Furthermore, samples are typically diluted with a non-absorbing reference powder to reduce both the sample's particle size dependence on the absorption coefficient and the amount of specular reflection relative to diffuse reflection. Specular reflection distorts the spectra of diffusely reflected light by modifying the shapes and relative intensities of the spectral bands. Dilution with a transparent reference allows the light to penetrate deeper into the sample and increases the amount of absorption and diffuse scattering.[12]

Just as with transmission spectroscopy, DRS features a very broad range of applications. The technique is frequently used to characterise materials such as paint, paper, textile, pharmaceuticals, ceramics, nanostructured surfaces, coal and soil powders, to perform surface and adsorption studies in organic, inorganic and physical chemistry,[13] to investigate a broad range of biological materials, and is even used in clinical studies *in situ* or using needle biopsies. DRS is also used to characterise heterogeneous catalysts and to provide information about the oxidation state, coordination geometry, degree of condensation, particle size and nature of a chemical bond.[14–16] The technique provides a more accurate method to determine the electronic structure and bandgap of semiconductor nanomaterials,[17,18] and circumvents the requirement of transmission spectroscopy that a powdered sample should dissolve well in a liquid.

3.5 Time-resolved Spectroscopy

3.5.1 *Introduction and equipment*

In Section 3.2.5, we explained that a fast detector allows one to perform time-dependent measurements. In fact, for a time resolution down to the microsecond time scale, the optical setup can be largely the same as for steady-state measurements (Figures 3.1 and 3.3). With such time resolution, spectroscopic changes of numerous types of reaction kinetics can be monitored in real time. In addition, phosphorescence of most materials and molecules occurs on a time scale of microseconds to milliseconds (see Section 2.5).

It is informative to measure the time dependence of the full spectrum. We will first briefly evaluate the current options for fast detectors. A single photodiode and PMT can be operated very fast and efficiently and also offer the cheapest solution; however, using such a point detector to measure time-dependent spectral changes results in either impractically long measuring times or a too severe limitation in the time resolution. Multichannel detectors circumvent this problem. Considering that a dispersive element such as a grating spreads a spectrum spatially out along a line, we require a detector with a 1-dimensional array of light-sensitive pixels to measure a full spectrum at once. CCD and CMOS cameras are

2-dimensional detectors but they generally have the option of changing the region of interest. Using these cameras in spectroscopic mode, whereby the region of interest consists of one row of pixels or at most a couple of rows, allows for fast measurements. In spectroscopic mode, CCD cameras easily attain a frame rate of several kHz, corresponding to sub-millisecond time resolutions (where a frame rate is defined as the number of times the full region of interest is measured per second). CMOS cameras are often ~100 times faster than CCD cameras and, when operated in spectroscopic mode, can often achieve MHz frame rates, which correspond to microsecond time resolutions. Alternatively, one could use a 1-dimensional detector, also called a line detector, such as the commonly used photodiode array, which provides a more cost-effective option than a scientific-grade camera. Photodiodes with ultrafast (sub-picosecond) response times are available, although the time resolution of these devices is limited by the much longer dead time, which accounts for the signal propagation in electric circuits. Using fast electronics, photodiode dead times of about 1 nanosecond (ns) are available on the market. This brings one to the time scale of typical fluorescence lifetimes.

It is not possible to detect a spectrum every nanosecond, because there will simply be too little light to resolve a spectrum. Fortunately, most processes are repeatable and the number of photons can be accumulated until a spectrum with a reasonable signal-to-noise ratio is obtained. Time-resolved measurements are generally performed by gating the detected photons into time bins. For example, all photons arriving between 4 and 5 ns after the excitation pulse will be binned together during a certain acquisition time.

A resolution of 1 ns is often not sufficient. For example, to accurately determine a fluorescence lifetime of 1 ns requires time steps of at least 0.1 ns. Better time resolutions can be achieved using one of two special types of cameras. The first is an Intensified CCD camera, which makes use of a very fast shutter to bring the resolution down to ~100 picoseconds (ps) or less. A streak camera can achieve a resolution of ~60 ps when used in single-shot mode and 1 ps or less when used in multi-shot mode.[1,2] Using multiple laser beams through a method known as fluorescence upconversion, the resolution can be pushed to sub-ps times. These methods are described in Section 3.5.2.

If the fluorescence decay is independent of wavelength, time-correlated single photon counting (TCSPC) is the method of choice to measure decays down to a time resolution of ~10 ps. TCSPC is described in Section 3.5.3 and contrasted to an alternative, phase-modulation method.

The abovementioned high-resolution methods are all for fluorescence applications. Fluorescence is a very sensitive experimental parameter, in contrast to absorption. It is also challenging to resolve small absorption changes using a single beam transmitted through or scattered by a sample. For high-resolution time-dependent absorption measurements, a second beam of light is used to monitor absorption changes after excitation with a laser pulse or flash of light. This technique, known as flash photolysis or pump-probe spectroscopy, is discussed in Section 3.5.4.

A photoinduced reaction is one that is induced by an excitation (i.e. absorption of a photon). Time-resolved absorption spectroscopy allows one to follow the absorption changes of such a reaction before steady-state — or energy equilibrium — is reached. A sufficiently short excitation pulse is required to initiate the photoreaction, and the time at which photon absorption occurs should be known. An excitation-fluorescence cycle represents a photoinduced process. Measurement of fluorescence lifetimes therefore requires a short excitation pulse as well as knowledge of its arrival time at the sample. The photon arrival times can be determined very accurately by synchronising the detector with the excitation source, a process called gating.

The most appropriate pulsed excitation light source depends largely on the application. Flash lamps, pulsed LEDs and pulsed lasers can be used. One can also use a continuous light source together with a fast shutter. An optical chopper can deliver pulse durations in the microsecond range, while electro-optic and acousto-optic modulators can act as sub-microsecond shutters, for example by periodically transmitting the incident light for only very brief periods. Flash lamps and LEDs are economic solutions and commercial options can deliver pulses as short as hundreds of ps. Pulsed lasers are more expensive but provide significantly better sensitivity, time resolution and shorter acquisition times. Numerous types of lasers exist with ps and sub-ps pulse durations. Diode lasers are a relatively inexpensive type of laser for this purpose.

3.5.2 *Time-resolved fluorescence spectroscopy*

Fluorescence is a valuable experimental property due to its sensitivity to the direct environment of a fluorophore. Changes in a fluorophore's chemical environment such as temperature, pH or interaction with another molecule are reflected in the fluorescence spectrum and especially in the fluorescence lifetime. In addition, processes such as energy transfer, charge transfer, quenching, rotational diffusion and solvation dynamics typically occur on the same time scale as fluorescence decay and can therefore be investigated through time-resolved fluorescence techniques.

Intensified CCD camera

In an intensified CCD camera, the detected signal is amplified inside an image intensifier tube. This method not only enhances the detection sensitivity to the single-photon level but, more importantly, also allows for very fast gating. Gating works as follows. The gated detector is synchronised with the pulsed excitation source and gating is done by means of voltage pulses. The user determines the gating periods before the measurement starts. The gating period is the time interval after an excitation pulse arrives at the sample during which the detector is sensitive. This means that only photons that hit the detector during the gating period are detected. In this way, a different 2-dimensional slice of the 3-dimensional time-dependent fluorescence decay spectrum is sampled during each gating period. Photons are accumulated during each gating period until a satisfactory signal-to-noise ratio is obtained.

Gating allows one to better distinguish between different decay signals in a sample, for example when the sample exhibits multicomponent fluorescence decay or a combination of fluorescence and phosphorescence. The short-lifetime component can then be gated out by using only gating periods corresponding to relatively long times after excitation. Similarly, scattering and other background signals, which arrive at the detector immediately after excitation, can be eliminated.

Recent technology has delivered intensified CCD cameras with sub-ns gating periods. The first generation of these ultra-fast shutter cameras provided a gating period of 200 ps, for which a time resolution of 25 ps was reported for time-resolved fluorescence spectral measurements.[19] More recent technologies allow for even higher resolutions, for example, LaVision's PicoStar UF camera offers a gating period of 80 ps and possibly <10 ps resolution.

Streak camera

A streak camera allows one to rapidly obtain a complete time-dependent fluorescence spectrum with ps time resolution. If the sample's fluorescence signal is sufficiently strong, the complete time and wavelength dependence of the fluorescence spectrum can be obtained with a single excitation pulse!

A streak camera is, in fact, much more than a camera. The light-sensitive area of the detector is a photocathode, which converts the incident photons into electrons. The electrons are then accelerated in an electric field and pass through a deflection system where the electrons are exposed to a rapidly increasing sweeping voltage. Electrons that arrive at a later time in the deflection system, corresponding to longer fluorescence decay times, are exposed to a higher voltage and are deflected more than electrons arriving earlier. The deflection system therefore converts the photon arrival time at the detector into a measurable position. After deflection, the electron beam is amplified, falls on a phosphor screen where the electrons are converted back into photons and finally detected by a sensitive CCD camera. This configuration allows for lifetime measurements only. To access spectral information, a spectrograph is added before the CCD camera to spectrally disperse the photons. All the different components comprising the streak camera multiply the cost of the device.

There are two basic types of streak cameras, depending on how the voltage sweeping is performed in the deflection unit. In a single-sweep device, the voltage is increased linearly, while the synchroscan system makes use of high-frequency (MHz–GHz) sine-wave modulated voltage patterns and accumulates data from multiple excitations. A single-sweep device can deliver a resolution down to ~60 ps and access a time window up to 10 ms,[1,2] allowing for time-resolved measurements of both fluorescence and phosphorescence spectra. Synchroscan cameras often bring the resolution below 1 ps. Hamamatsu, the market leader, boasts with its FESCA-100 femtosecond (fs) streak camera featuring a time resolution of 150 fs or less.

Higher-resolution systems not only dramatically increase the cost of the device but are also accompanied with a larger amount of jitter in the signal that triggers the voltage sweeping. This trigger jitter results in changes in the amount of deflection of the electrons passing through the deflection system and thus limits the time resolution. A single-pulse measurement is not influenced by this jitter, but as soon as averaging is required the trigger jitter determines the time resolution of the synchroscan system. The best compromise between high-frequency sweeping and trigger jitter gives a time resolution of about 1 ps.[20]

Fluorescence upconversion

Instead of using electronics to gate the fluorescence signal, like in the case of intensified CCD cameras, light pulses can also be used. This procedure is called optical gating and is the basic principle of fluorescence upconversion. In fluorescence upconversion, the fluorescence signal is overlapped with a second beam inside a crystal. The crystal is a specific, nonlinear medium in which the two beams can interact to produce a third beam of a different colour by means of upconversion (see Section 7.4.6), which is dispersed and finally measured. The second beam is the gating beam. The exact instant at which a gating pulse overlaps with the fluorescence beam inside the crystal determines which part of the long fluorescence pulse is sliced out. The timing between the two beams is finely controlled by increasing or decreasing the path length of the gating pulse in small increments. The time resolution of the measurement is essentially determined by the duration of the gating pulse, which can be as short as tens of femtoseconds.

Another optical gating technique for sub-ps time resolution is based on the Kerr effect, which will not be discussed here. This technique offers the benefit of measuring the original fluorescence beam after optical gating via a Kerr shutter, instead of an indirect measurement via a third beam like in the case of fluorescence upconversion.

It is important to note that experimental techniques based on optical slicing are expensive and not simple to perform, as they require very precise optical alignment, along with expensive ultrafast lasers, high-quality optical components, and fast and sensitive detectors.

3.5.3 *Fluorescence and phosphorescence lifetime measurements*

The fluorescence lifetime of a sample can be measured either in the time or in the frequency domain. In the time domain,

TCSPC is the method of choice due to its superior sensitivity and simpler data collection and analysis compared to other methods. This is a direct method for measuring the lifetime and makes use of pulse fluorometry. In the frequency domain, phase fluorometry is applied, which is an indirect method, because the lifetime cannot be directly obtained from the measurement but should be calculated via a Fourier transform, assuming a particular shape of the fluorescence decay. TCSPC provides a time resolution from ~10 ps to ~100 ns, offering an ideal method for fluorescence lifetime measurements, while phase fluorometry can resolve changes in the measured signal on time scales ranging between ~1 ns to ~1 ms and is therefore more applicable to phosphorescence lifetimes. Nowadays, both pulse and phase fluorometry are routine methods and user-friendly commercial instruments with analysis software are available. The basic principles of both methods will be described here.

Pulse fluorometry (time-correlated single photon counting, TCSPC)

When a population of identical fluorophores are excited at the same time, the fluorescence intensity F after a time t generally follows the behaviour

$$F(t) = F_a e^{-t/\tau}, \tag{3.4}$$

where the amplitude F_a is the intensity immediately after excitation, defined as time $t = 0$. This equation represents a single-exponential decay function, which indicates that the largest fraction of fluorophores fluoresce immediately after excitation, and this fraction decreases to $1/e$ after the time $t = \tau$. This means that 63.2% of the fluorophores have fluoresced after the time τ. The constant τ is the fluorescence lifetime, also known as the fluorescence decay time, and is exactly equal to the average time the fluorophores remain in an excited state.

Measuring the fluorescence intensity of a fluorophore population as a function of the time after excitation gives a fluorescence decay trace (Figure 3.6(a)).

In pulse fluorometry, it is important to realise that Eq. (3.4)
represents a probability distribution. This means that when a
single fluorophore is excited and does not transfer its energy to
another molecule, the probability that the fluorophore emits a
photon via fluorescence at a time t after excitation is also given
by this equation. In other words, after a sufficient number of
excitation-fluorescence cycles of a single fluorophore, the fluorescence decay curve in Figure 3.6(a) can be reconstructed. This
is done by grouping the arrival times of the detected fluorescence photons in short time intervals (called time bins), giving
a fluorescence decay histogram, which is built photon by photon. This is the principle of TCSPC.

A TCSPC trace is constructed from emission of a single
fluorophore at a time. This does not mean that only a single
fluorophore should be used. The technique is, in fact, sensitive

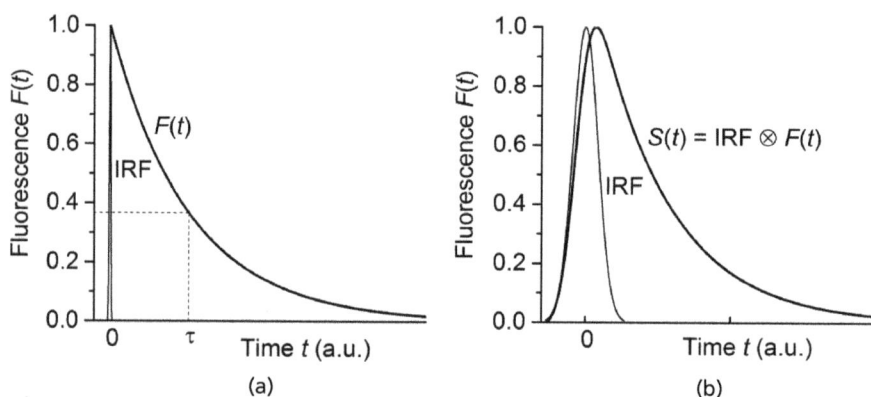

Figure 3.6: Mono-exponential fluorescence decay traces, as described by
Eq. (3.4). The instrument response function (IRF) is indicated by the pulse at
time 0, and τ is the fluorescence lifetime. (a) IRF significantly shorter than τ
and (b) IRF comparable to τ, so that the measured response function $S(t)$ is
the convolution between the IRF and $F(t)$ and calculated from Eq. (3.15).
All pulses are normalised.

enough to be applied to a single fluorophore. This application will be discussed in Section 7.1. However, single fluorophores exhibit a phenomenon called blinking, whereby the fluorescence lifetime suddenly changes. They also photobleach after a finite number of excitation-fluorescence cycles. By using a population of identical fluorophores, these two problems are overcome and fluorescence decay curves of much higher quality are obtained. For such a population there are two necessary conditions that have to be met: (i) the fluorophores should all be excited simultaneously, and (ii) the arrival times of the fluorescence photons should be discriminable by the detector. The first condition is met by using a single, short excitation pulse, such as shown in Figure 3.6(a). To meet the second condition an important technical limitation should be taken into account: standard electronics in TCSPC devices have a typical dead time on the order of tens of ns. The shortest dead time offered by current devices is approximately 1 ns, which is much longer than the required time resolution for fluorescence lifetime measurements. During the dead time no photon can detected. This means that when a stream of photons hits the detector, in most cases only the first photon, corresponding to the shortest lifetime, will be detected. This phenomenon is known as the pile-up effect and results in an artificial shift of the decay curve to shorter times. To avoid the pile-up effect, at most one photon should be detected for every excitation pulse. This is achieved by using a sufficiently weak excitation intensity as well as a time delay between excitation pulses significantly longer than the fluorescence lifetime of the sample. The latter condition ensures that a photon is detected before the next excitation pulse arrives.

The exact time between excitation and photon emission is determined by comparing the arrival time of a photon at the detector with a reference clock. The reference signal is the periodic excitation pulse. The time resolution of a TCSPC setup is called the IRF and is determined by the combination of the excitation pulse duration and the detector's response time and timing resolution. The detector's timing resolution is the time taken by the detector to convert the photon into an electronic signal, and generally contributes the largest amount of broadening to the IRF. The convolution of the IRF with the emission function is mathematically described by Eq. (3.15).

It is important to note that the IRF is temperature dependent and may slowly change when the detector's temperature changes. It is therefore recommendable to measure the IRF before every measurement. The IRF of a TCSPC setup is obtained by measuring the TCSPC trace of a reference sample. There are three types of reference samples that can be used:

1. A scattering solution such as Ludox (colloidal silica), glycogen or coffee creamer.
2. A strongly quenched fluorophore. It was demonstrated that strong quenching can be induced on numerous fluorophores using a high concentration of potassium iodide.[21,22]
3. A fluorophore with a very short intrinsic fluorescence lifetime, such as pinacyanol iodide in methanol (lifetime of 6 ps),[23] LDS 798 in water (27 ps lifetime),[22] and pyridine in water (37 ps lifetime).[22]

Equation (3.4) is only valid when the sample contains a high purity of a single type of fluorophore and the fluorophores do not transfer their energy to other molecules. In the case of energy transfer or quenching, the fluorescence decay has a non-exponential form. A mixture of different, non-interacting fluorophores produces a multi-exponential TCSPC trace and their respective lifetimes can be resolved unambiguously if only a small number of different fluorophore types are present and their lifetimes are well separated (see Section 3.7.3). Rotational diffusion also gives rise to multi-exponential fluorescence decay if the diffusion time is comparable to the fluorescence lifetime. This is the case for sufficiently large samples, such as fluorophores bound to a protein. The influence of rotational diffusion on the fluorescence decay can be suppressed by using a linear polariser before and after the sample such that their polarisation axes are offset by the magic angle (see Section 3.6).

When spectral information is not required, TCSPC offers several advantages compared to other methods:

1. A sufficiently high resolution for most fluorescence lifetime studies can be obtained. Microchannel plate PMTs offer IRFs as short as 30–40 ps, from which decay times of 10–20 ps can be resolved when performing proper deconvolution during data analysis (see Section 3.7.3).
2. Highly accurate measurements can be performed on samples with very low fluorescence intensity. Samples can be diluted even to the level of a single fluorophore.
3. TCSPC is performed in the time domain, which makes interpretation of the data easier. For example, the fluorescence decay trace represents the probability that the sample is in an excited state for all measured times after excitation.
4. The fluorescence decay trace obtained from TCSPC is insensitive to the stability of the light source.
5. The data is cumulative. In other words, provided the sample remains stable, the user can continue data collection until a satisfactory signal-to-noise level is achieved.
6. Compared to other time-resolved spectroscopy methods, TCSPC is a cost-effective option.

The major disadvantages of TCSPC are the long measuring times required for high-quality data, the limited time resolution compared to techniques based on optical gating (see Section 3.5.2), and the fact that this technique does not provide spectral information. The last limitation can be partially overcome by repeating the measurement several times, each time using another band-pass filter in the emission path to measure the fluorescence decay in various spectral windows sequentially. Alternatively, parallel detection is possible by dividing the fluorescence amongst a number of detectors, each measuring a different spectral window. Whereas the first solution significantly prolongs the time of measurement, the second solution may considerably increase the cost of the instrument.

Phase fluorometry

Since frequency and time are interconvertible through Fourier transformations, the fluorescence or phosphorescence emission lifetime τ can also be determined from measurements performed in the frequency domain. In the frequency domain, one is interested in changes in particular parameters related to τ that are measured as a function of frequency instead of as a function of time. In phase fluorometry this is done by modulating the intensity of the excitation light at a frequency comparable to the time scale of the emission. For example, if a fluorescence lifetime of 10 ns is to be resolved, the modulation frequency should be in the order of 10–100 MHz. The emission intensity will follow the same modulation pattern, which will be offset from that of the excitation light due to the delay between excitation and emission. The time delay between the excitation and emission modulation patterns therefore reflects τ. A sinusoidal modulation function is typically used but the modulation can also be in the form of discrete light pulses. The delay between the excitation and emission signals is known as the phase shift φ, and an average τ can be calculated directly from φ.

If the modulation frequency is too high, or, equivalently, the excitation pulses are too short, only very small values of φ are obtained. This is the case for Figure 3.6(a), where the IRF is considerably shorter than τ, so that the measured fluorescence decay curve is hardly affected by the IRF and φ is almost zero and virtually independent of τ. However, when the IRF is of the same order of magnitude as τ, the measured fluorescence or phosphorescence response curve is given by the convolution between the IRF and the actual emission signal, as shown in Figure 3.6(b), and has a maximum after time zero. Figure 3.7(a) shows an example of such an IRF and a few representative emission signals corresponding to different lifetimes.

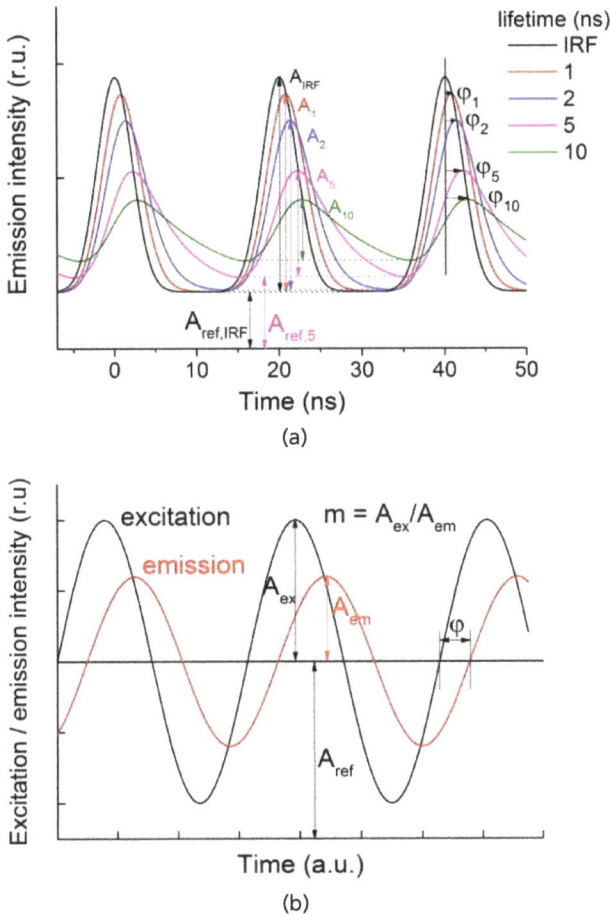

Figure 3.7: (a) Emission modulation patterns following pulsed excitation and (b) sinusoidally modulated excitation. The emission decay traces in (a) were calculated using Eqs. (3.4) and (3.15) and a Gaussian IRF with FWHM of 5 ns. Phase shifts and amplitudes are denoted by φ and A, respectively, where A_{ref} refers to the amplitude with respect to a reference intensity, shown in (a) for the IRF and the $\tau = 5$ ns decay traces.

As τ increases, φ also increases. Since the intensity under the decay curve should stay constant, the amplitude of the emission signal decreases as τ increases. The ratio between the amplitude of the emission signal and the amplitude of the IRF is known as the demodulation factor, m. This factor is a second parameter that can be used to determine τ. A similar behaviour is obtained when using a sinusoidal modulation pattern (Figure 3.7(b)). When the modulated emission and excitation signals have different baselines, like in Figure 3.8(a), their amplitudes are commonly scaled by their vertical offset with respect to a reference intensity.

The lifetimes associated with φ and m are called the phase and modulation lifetimes, represented by the symbols τ_φ and τ_m, respectively, and are calculated through the equations

$$\tau_\varphi = \frac{1}{\omega}\tan\varphi \text{ and } \tau_m = \frac{1}{\omega}\sqrt{\frac{1}{m^2} - 1}, \qquad (3.5)$$

where ω is the angular frequency, i.e. $\omega = 2\pi f$, with f being the modulation frequency. The accuracy of determining τ_φ and τ_m is improved by using a set of different values for f.

Comparison of the two lifetimes provides information about the shape of the emission decay. When the two lifetimes are the same, single exponential decay is involved, which means that the sample constitutes a single emitter and the two lifetimes are the real emission lifetime. When the two lifetimes are different, a more complex decay is involved and the lifetimes represent only approximate values of the emission lifetime. When $\tau_\varphi < \tau_m$, the emission exhibits multi-exponential decay, likely due to a mixture of emitters. Appropriate weighting should be applied to determine the different lifetimes involved in multi-exponential decay. By analysing the differences in the response of τ_φ and τ_m to the modulation frequency, an indication of the shape of the emission decay can be obtained.

Phase fluorometry is more suitable for measurements of phosphorescence and long fluorescence lifetimes, as the technique is prone to large uncertainties when short lifetimes are resolved.[24] Furthermore, the range of modulation frequencies should be chosen such that φ is not too large, ideally <70°, otherwise the phase lifetime is again very sensitive to uncertainties in φ.[24]

Although TCSPC is often the preferred technique for emission lifetime measurements, phase fluorometry offers a fast and inexpensive technique and is particularly useful when a large number of samples with emission lifetimes of >1 ns have to be measured. The technique is superior to TCSPC on long time scales and is therefore an ideal method for determining phosphorescence lifetimes. An additional advantage is that no deconvolution of the IRF and emission signal is necessary like for TCSPC, because the data is analysed directly in the frequency domain. In fact, the phase shift and demodulation parameters are a direct result of the IRF. Finally, phase fluorometry can be integrated quite easily into a steady-state spectrofluorometer by adding three key elements: (i) an electro- or acousto-optic modulator to modulate the excitation light when using a continuous light source (if a pulsed light source with adequately short pulses is already included, a modulator is not necessary); (ii) a second, cheap detector such as a photodiode to supply the reference signal from the excitation pulse (many spectrofluorometers already include such a detector); and (iii) a lock-in amplifier to synchronise the excitation and emission signals and thereby very precisely detect the phase shift between the two signals.

3.5.4 *Time-resolved absorption spectroscopy*

The two-pulse technique known as pump-probe spectroscopy or *flash photolysis* makes it possible to investigate photophysical and photochemical processes on any time scale. The technique was developed by Ronald Norrish and George Porter in the late 1940s to resolve reaction kinetics on the microsecond time scale, an achievement that was awarded the Nobel Prize in 1967. Upon invention of the laser in the early 1960s, nanosecond pulses became available and even faster light-induced kinetic processes could be monitored. Thereafter, the time resolution of pump-probe spectroscopy rapidly advanced with the technological development of the laser, from nanoseconds to picoseconds to femtoseconds. This resolution provides the means to follow the course of the fastest processes in atoms and molecules, such as molecular vibrations. For example, Ahmed Zewail has pioneered the field of femtochemistry by using laser pulses as short as tens of femtoseconds to study the nature of chemical bond formation and chemical reactions. This work led to his Nobel Prize in 1999.

Pump-probe spectroscopy makes use of two pulses, known as the "pump" and "probe" pulses. The pump pulse arrives at the sample first and excites a sub-population to an electronically excited state or triggers the start of a chemical reaction. The probe pulse arrives at the sample at a specified delay time after the pump pulse and its absorption spectrum is measured. The measurement is repeated a large number of times, every time increasing the delay between the pump and probe pulses. In this way the photoinduced course of events can be followed from its initiation at time zero, where the pump and probe pulses arrive at the sample simultaneously, until the reaction product is formed, energy is transferred to the acceptor or equilibrated in the system, or the excited state has relaxed back to the ground state. Since the excited state, all intermediate states, and the product states have unique absorption spectra, a different absorption spectrum of the probe beam is measured at every delay time.

With such a sensitive experimental tool giving access to femtosecond to picosecond time scales it is possible to follow molecular photophysical processes (such as internal conversion, vibrational relaxation and energy transfer), reactions that involve proton or electron transfer, collisional dynamics, molecular vibrations, isomerisation and the chain of events following important photoinduced reactions in biology, such as vision and photosynthesis. Reaction intermediates like radicals, ions and triplet excited states can be identified, and the rates of all abovementioned processes can be determined. Most intermediate states are short-lived and are therefore referred to as transient states, giving the technique also the name of transient-absorption spectroscopy.

The pump pulse is an intense pulse to excite a sizeable population of the sample, ideally more than 10%, but the intensity should still be low enough to not damage the sample or induce absorption of more than one photon per molecule or complex, unless the purpose of the experiment is to investigate multiphoton processes. The pump beam typically covers a spectral window as narrow as possible to target a specific electronic transition. The duration of the pump pulse, together with the IRF, determine the time resolution of the experiment. According to Heisenberg's uncertainty principle, the wavelength resolution scales inversely with the time resolution. This limitation becomes important when using sub-picosecond pulses, because the narrowest bandwidth of such pulses covers several wavelengths and therefore excite a broad band of vibrational states.

The probe beam typically covers a large wavelength window in order to extract as much information as possible from its absorption spectrum measured after its interaction with the sample. Its spectrum should ideally cover the full absorption spectrum of the sample's intermediate and product states. The probe pulses should have a sufficiently low intensity, so that multiphoton processes are completely avoided.

In pump-probe spectroscopy, one is interested in changes in the absorption spectrum at different delay times, which means that all absorption spectra have to be compared directly with the steady-state absorption spectrum. Simply subtracting the steady-state absorption spectrum from all transient-absorption spectra after the measurement will give a result of poor quality. This is because the changes in the spectral amplitudes are in the order of 1–10 mOD, which is 100–1,000 smaller than the band maxima of the steady-state spectrum, while the intensity of the pump and especially the probe beam fluctuates, and the full extent of intensity fluctuations over the course of the measurement is often also in the mOD range, even for state-of-the-art pump-probe setups, and especially for picosecond and femtosecond instruments. Since the intensity variations are most significant on a time scale of seconds to minutes, their effect is limited when the steady-state absorption spectrum of the sample is measured a few milliseconds after each transient spectrum. This is done by blocking every other pump pulse before it reaches the sample but measuring every probe pulse. When a pump pulse is blocked, the probe pulse gives the steady-state spectrum, which is matched with the transient spectrum of the previous probe pulse. The pair of spectra are subtracted from each other, giving a difference absorption spectrum corresponding to each delay time. In other words, if A_{pumped} and $A_{unpumped}$ correspond to the measured absorbance at a particular wavelength when the sample was excited by the pump pulse and when the pump pulse was blocked, respectively, the difference absorption is given by the following expression:

$$\Delta A = A_{pumped} - A_{unpumped} = \log\frac{I_{unpumped}}{I_{pumped}}, \qquad (3.6)$$

where the second part of the equation follows directly from the law of Lambert–Beer. ΔA is a function of both wavelength (λ) and delay time (τ).

Due to the small size of ΔA, typically ranging between 1 and 20 mOD, it is important to optimise the measured signal and to limit the noise as efficiently as possible. For this purpose, an optical density of 4–6 of the sample should ideally be used, the spatial overlap between the pump and probe beams should be optimised, and the probe intensity should be optimal. An optimal probe intensity gives a measured signal that corresponds to roughly 50% of the detector's dynamic range and is typically obtained by focussing the probe beam tightly into the sample. In fact, by focussing both beams tightly, the noise level of the detected signal is also reduced and the time resolution is improved. Consider for example that a pump and a probe beam, both with a 1 mm diameter, are used. In most setups the beams are not collinear, which means they enter the sample at a small angle with respect to each other. This is done to limit the detection of scattering of the pump beam. For illustration, let us assume an angle of 90° between the two beam paths. The region of overlap of the two beams is then a square with sides of 1 mm. It takes light 3.3 ps to traverse a distance of 1 mm. The probe beam therefore takes 3.3 ps to travel through the wave front of the pump beam. The wave front is the front end of the beam where all points have the same phase and thus interact with the sample at the same time and in the same way. This means that the molecules that are excited simultaneously by the pump beam, "see" the probe beam over the entire duration of 3.3 ps. If the pulses have femtosecond duration, the time resolution of the experiment is drastically decreased to 3.3 ps.

It is also important that the focal spot size of the probe beam is smaller than that of the pump beam, so that only pumped molecules are probed. Finally, the spectrum of the probe beam should be as flat as possible, i.e. its intensity should be relatively independent of wavelength, because peaks or sharp changes in the spectrum are often imprinted on the difference absorption spectrum, leading to artefacts and increased noise.

While time-resolved fluorescence techniques only provide information about transitions between the lowest optically-allowed electronic excited state and the ground state, pump-probe spectroscopy additionally gives spectroscopic signatures of non-emissive states as well as of vibrational relaxation of the excited states and the ground state. For example, according to the Franck–Condon principle, excited-state decay via fluorescence emission takes place typically to a higher vibrational level of the electronic ground state, after which ultrafast relaxation occurs to the lowest vibrational level. A fluorescence-based technique can monitor only the first step, while transient absorption spectroscopy can monitor the last step too, because different vibrational levels have different absorption spectra. However, transient absorption spectra are more difficult to interpret than time-resolved fluorescence spectra. When fluorescence or phosphorescence is one of the processes that take place, time-resolved fluorescence or phosphorescence spectroscopy can be used as a complementary technique to separate the dynamics related to fluorescence or phosphorescence from the other signals.

3.6 Polarisation-dependent Spectroscopy

It is relatively simple and inexpensive to perform polarisation studies using a standard spectroscopy instrument. By adding one or two polarisers, a wide range of applications are added. Besides orientational information of the chromophores, information can be obtained about the surface topology, the local viscosity of the solvent, membrane fluidity, conformational changes of proteins, configurational changes of chromophores, association reactions between biological molecules, denaturation of proteins and DNA, rotational diffusion of molecules, and energy transfer between molecules.

The polarisation dependence of absorption can be studied by placing a polariser in a spectrophotometer before the sample and compare the absorption spectrum for two perpendicular polarisations of the exciting light, typically horizontally and vertically polarised light. For fluorescence measurements two polarisers are used to polarise the exciting light and emission light independently. By polarising each either horizontally or vertically gives four polarisation combinations. The two polarisers are typically placed just before and just after the sample, respectively. It is common practice to use only vertically polarised exciting light and compare the horizontal and vertical polarisations of the emission spectrum. Horizontally polarised exciting light is then used to correct the spectra due to the polarisation dependence of the optical components before the sample, usually the light source and grating of the excitation monochromator.

For steady-state absorption measurements, a polarisation dependence will only be observed when the chromophores are partially or completely aligned, for example in a liquid crystal matrix (Figure 4.11). A preferential orientation can be induced by placing the chromophores in an external electric field or by immobilising them in a gel matrix and compressing the gel along one axis.

For time-resolved absorption measurements, one is interested in polarisation changes that occur during the time between the pump and the probe pulses. The polarisation and anisotropy are calculated using Eqs. (2.11) and (2.12) but replacing $I_{\parallel}$ and $I_{\perp}$ with $\Delta A_{\parallel}$ and $\Delta A_{\perp}$, respectively, where ΔA is the absorbance difference defined by Eq. (3.6), and $\Delta A_{\parallel}$ and $\Delta A_{\perp}$ correspond to arrangements where the pump and probe beams are polarised parallel and perpendicularly to each other, respectively. Polarisation changes are assessed even if the molecules are randomly oriented, because a preferential orientation is selected by the pump beam. For fluorescence measurements, the exciting light similarly selects molecules having a preferred orientation and no additional alignment of the chromophores is necessary.

A few important measures have to be taken to ensure accurate polarisation measurements:

- Several optical units are polarisation dependent. The diffraction grating often shows the strongest dependence, i.e. its efficiency depends strongly on the polarisation of the incident light. The light after a grating is therefore partially polarised, even if the incident light is unpolarised. As a result, changing the orientation of a polariser after the grating changes the transmitted light intensity. This has to compensated for in the measured spectra. For absorption spectra, the intensity of the exciting light can simply be scaled accordingly so that the light arriving at the sample has the same intensity for both polarisations. For the measurement of emission spectra, the intensity transmitted by both the exciting and emission monochromators depends on the polarisation. The former is generally corrected for by measuring the polarised emission spectra using horizontally polarised exciting light instead of vertically polarised exciting light. The polarisation dependence of the optical units after the sample (typically the emission monochromator and detector) is given by the G factor appearing in Eqs. (2.11) and (2.12). To calculate the G factor, the emission spectra are measured after placing the two polarisers such that their respective transmission axes are at an angle of 54.7°, known as the magic angle, and compare the spectra with those obtained in the usual configuration (90° and 0° between the polarisation planes of the exciting and emission light). The magic angle orientation ensures that the total measured intensity is independent of the polarisation.

- Light scattered perpendicularly to the direction of the incident beam is completely polarised, even if the incident light is unpolarised. This is the case for fluorescence spectroscopy based on the right-angle configuration (Figure 3.3). Such scatter may strongly distort the polarisation spectra and is critical to be eliminated.

- It is important to know the polarisation state of the light source. For example, halogen lamps produce unpolarised light while the light output from most lasers is plane polarised. Waveplates can be used to rotate the polarisation plane without attenuating the intensity. For example, if a laser is used that produces horizontally polarised light and vertically polarised light is required, the beam can be passed through a half-wave plate to rotate the polarisation plane through 90°. The polarisation plane can be rotated through any arbitrary angle without loss of the beam's intensity by using two quarter-wave plates in series. Due to the wavelength dependence of waveplates they cannot accurately rotate the polarisation plane of a full spectrum at once but smaller wavelength intervals can be used sequentially.

3.7 Data Analysis Methods

3.7.1 *Introduction*

The objective of data analysis is to realistically interpret the experimental data. Data analysis always involves the use of one or more mathematical models (i.e. a mathematical equation) and a particular method to use this model. The goal is to find a mechanistic model, which is a model that gives physical meaning to the relationship or classification that is identified. The information obtained from a mechanistic model describes the properties of a real system, such as rate constants or diffusion constants. The model is empirical when a mathematical equation is used to describe a particular relationship but the parameters derived from the model do not have a physical meaning. An example of an empirical model is a polynomial function that describes the baseline of a spectrum. Baseline correction is essential for realistic data analysis but does not correspond to any chemical or physical property of the studied system.

Most processes in the world are too complex to model completely using a mechanistic model, or a realistic model would be too complex to solve. The data analysis model is then simplified by considering only the most important subunits of the system and/or related parts of the system are grouped together.

3.7.2 *Goodness of fit*

In order to perform data fitting (often called regression) an appropriate predetermined physical model is needed. The physical model is described by a mathematical equation, for example a linear or quadratic function, or a Gaussian or Lorentzian distribution. Data fitting allows one to find the best variables of the model that describe the experimental data or the relationship between certain parameters.

The physical model is frequently assumed to be the simplest model that optimises the R^2 value or the χ^2 value, defined by

Eqs. (3.7) and (3.8). However, this is not necessarily a true assumption. The R^2 and χ^2 values are useful indicators of the quality of the fit when there is no doubt that the chosen mathematical function represents the best mechanistic model but these values should not be used to determine the best mathematical function. Instead, the fitting residuals should be inspected by eye. The residuals are the difference between the raw data and the fit and should reflect pure noise. If the residuals show some structure it means that the mathematical model used to fit the data is not the best one (or the regression was not executed correctly). The presence of structure can be inspected visually, though it is advisable to additionally make use of an autocorrelation function for an objective approach, because structure in the residuals can be subtle. For this purpose, the Durbin–Watson parameter is frequently used.

The R^2 value, also known as the coefficient of determination, is given by the following equation:

$$R^2 = \frac{\sum_{i=1}^{N}(\hat{y}_i - \bar{y})^2}{\sum_{i=1}^{N}(\hat{y}_i - \bar{y})^2}, \tag{3.7}$$

where N is the number of data points, y_i is the measured data value for point i, $\hat{y}_i$ is the corresponding calculated value obtained from the fit, and $\bar{y}$ is the average of all y_i. Note that y_i is the dependent parameter, for example the fluorescence intensity at a time t. Equation (3.7) is the regression sum of squares divided by the total sum of squares, where the regression sum of squares describes how much the regression curve differs from the average value of all data, and the total sum of squares describes how much the measured data points vary around their average value.

The χ^2 value is defined as the weighted sum of the differences between the measured (y_i) and calculated $(\hat{y}_i)$ values:

$$\chi^2 = \sum_{i=1}^{N}\left(\frac{\hat{y}_i - y_i}{\sigma_i}\right)^2, \tag{3.8}$$

where σ_i, the standard deviation for y_i, is the weighting factor, giving data points with small errors more weight. The reduced χ^2 value, defined as χ^2 per degree of freedom, is more commonly used to determine the goodness of fit. The degrees of freedom are the number of data points minus the number of variable fit parameters. A regression giving a reduced χ^2 value near unity for acceptable values of σ_i is considered a good fit.

3.7.3 *Regression with an assumed function*

Least-squares analysis

The best fit is the one that minimises the difference between the fit and the experimental data. The most commonly used regression method is known as least-squares analysis. This approach is equivalent to minimising χ^2, although the weighting factor in Eq. (3.8) is often excluded. The curve fitting is performed iteratively, which means that different values for the parameters are tried and the values are improved in each iteration. All iterative processes require starting values for the parameters. When the physical model is based on a nonlinear function, commercial curve-fitting software may not always find appropriate starting values, and the fit may fail to converge if the user does not provide good estimates.

As an illustration, let us consider the time-dependent fluorescence decay of a molecular dye. One typically starts with the simplest model, which is described by a single exponential decay function such as given in Eq. (3.4). If the residuals have structure, another exponential is added. This process is repeated until no structure remains in the residuals. However, it is not recommended to use more than three exponentials, otherwise the data may be overfit, resulting in artefacts. The limited number of exponentials that adequately fits the data may indicate that the associated lifetimes are only averages and caution should therefore be taken to assign a physical value to them if it is known that more than three lifetimes may be present.

The time-dependent decay of the fluorescence intensity as a result of N simultaneous decay processes is described by

$$F(t) = \sum_{j=1}^{N} \alpha_j e^{-\frac{t}{\tau_j}},$$ (3.9)

where α_j is the fractional amplitude of each component j with corresponding lifetime τ_j.

Equation (3.9) represents a continuous function, but in reality the measured fluorescence counts are integrated into time bins. Suppose that we use m time points. Then, for each time point i, we write the associated fluorescence intensity as follows:

$$F(t_i) = \sum_{j=1}^{N} \alpha_j e^{-\frac{t_i}{\tau_j}},$$ (3.10)

where $i = 1, \ldots, m$. To express the full model it is more convenient to use matrix notation.[25]

$$\mathbf{f} = \mathbf{Ta},$$ (3.11)

where $\mathbf{f}$ is a vector of size m, containing all $F(t_i)$ values, $\mathbf{a}$ is a vector of size N, containing all amplitudes α_j, and $\mathbf{T}$ is the decay matrix of size $m \times N$, with components

$$T_{ij} = e^{-\frac{t_i}{\tau_j}}.$$ (3.12)

When regression is performed using a polynomial function it is referred to as linear regression; otherwise the regression is nonlinear. It is mathematically simpler to linearise a non-polynomial function and then perform linear regression than to do nonlinear regression. A power-law function can be transformed to a linear function in a straightforward manner by taking the logarithm. This would be the preferred method for the abovementioned example of fluorescence decay traces, where exponential functions are involved. All other non-polynomial functions can be linearised using a Taylor expansion around a selected value of the independent parameter (such as time). This results in a power series, which is a polynomial function, and hence linear regression can be performed. Our aim is therefore to obtain a multiple linear regression model, which can be written in matrix form as:

$$\mathbf{y} = \mathbf{Xa}. \tag{3.13}$$

In this expression, $\mathbf{y}$ is a vector containing m dependent target variables (such as the calculated fluorescence intensity in each time bin), $\mathbf{a}$ is a vector of size N containing all regression coefficients, which are to be determined through regression, and $\mathbf{X}$ is an $m \times N$ matrix containing the values of polynomial expressions of the independent variables (for example time) such that their matrix multiplication with the regression coefficients gives the components of $\mathbf{y}$. In other words, the calculated values of $\mathbf{y}$ are expressed as linear combinations of the polynomial terms, scaled by the regression coefficients. In Eq. (3.13), the independent variables are now completely separated from the regression coefficients.

Least-squares regression is the most common approach to data analysis. The attractive aspects of this method include its speed, reliability, versatility and the use of a mechanistic model, which lends to physical interpretation of the data. However, the assumption of a mathematical model prior to data fitting is also the major limitation of this technique. Furthermore, if error bars are assumed symmetric before linearisation of a function they no longer are after this transformation. This complicates least-squares analysis and error propagation. Finally, ordinary least-squares regression is only valid when the experimental errors have a Gaussian distribution. Fluorescence lifetime data, which features Poisson noise, is an example where this method will not be accurate.

Maximum likelihood estimation

One of the pitfalls of the standard least-squares approach is that the regression can converge to a local minimum and may never reach the required goodness-of-fit condition such as a certain threshold value for R^2 or χ^2. This may be particularly problematic in the case of noisy data. Another potential problem is that there may be different sets of parameters for the chosen mathematical fitting function that give the required goodness-of-fit condition. Different sets of starting values may give different solutions and it is not guaranteed that the most optimal fit will be found. These problems are overcome by maximum likelihood estimation. This is a probabilistic method but may also be considered an extension of the least-squares approach. In this method, all possible values of the parameters of the chosen mathematical fitting function are investigated, which guarantees that an optimal fit will be obtained. Furthermore, this method can accurately deal with experimental data having any noise distribution.

Global analysis

Experimental data sets that are related by some quantity ought to be analysed simultaneously. Such type of analysis is known as global analysis. The quantity that links the different data sets can be the absorption or fluorescence wavelength, temperature, pressure, or any other experimental variable. A well-known example is time-resolved spectroscopy data, whereby different processes are linked by the wavelength. For example, energy transfer between a donor and acceptor will appear as a decrease in the signal(s) related to the donor and an increase in the acceptor's transient spectrum. The mathematical equations describing the rate of decay and rate of formation, respectively, are linked through the wavelength.

To analyse time-resolved spectroscopy data, the goal is to calculate the transient spectrum $\Psi(t,\lambda)$, which is a function of both time (t) and wavelength (λ), that best describes the experimental data. Since $\Psi(t,\lambda)$ is a superposition of different spectral components, the model needs to decompose $\Psi(t,\lambda)$ into all components. If N spectral components contribute to the spectrum, the calculated transient spectrum is given by

$$\Psi(\lambda,t) = \sum_{j=1}^{N} c_j(t)\varepsilon_j(\lambda),$$

(3.14)

where $\varepsilon_j(\lambda)$ is the spectrum of the jth component and $c_j(t)$ is its amplitude at time t. To avoid overfitting the data, N is kept as small as possible. N is typically determined by inspecting the singular value decomposition of Ψ, a matrix factorisation technique to estimate the number of linearly independent components.

Deconvolution

When the time resolution of an experimental technique is limited by the IRF, the measured data is convoluted with the IRF and deconvolution should be performed to establish the real behaviour of the sample. Deconvolution is therefore critical for most time-resolved experimental data. We will use TCSPC again as an example. Equation (3.10) describes the response of the sample to an infinitely short excitation pulse but does not represent the measured behaviour. The measured data is affected by the response of the detector, given by the IRF. The measured signal S at time t is then given by the convolution between the IRF with the function $F(t)$:

$$S(t) = \text{IRF} \otimes F(t),$$

$$= \int_0^t \text{IRF}(t')F(t-t')dt',$$

(3.15)

where the symbol ⊗ signifies convolution. The effect of the IRF can be neglected only when it is considerably shorter than the lifetime. In such a case, the initial part of a TCSPC trace, which corresponds to the IRF, can be excluded from the fit and no deconvolution is necessary. However, in most cases, the IRF cannot be removed from the data but deconvolution should be done during the regression by considering the IRF as an additional function that affects every data point, i.e. the model function $F(t)$ should be multiplied with the IRF during data fitting.

Similarly, for global analysis, the measured signal is the convolution between the IRF and the calculated transient spectrum at every time point. Convolution is accounted for during regression by multiplying $c_j(t)$ in Eq. (3.14) with the corresponding IRF at that time point.

Fitting of fluorescence spectra

Inhomogeneous broadening of absorption and emission spectra usually dominates homogeneous broadening. Doppler broadening is a prominent component of inhomogeneous broadening and results from the speed distribution of molecules in a solution. Another contribution to inhomogeneous broadening results from the interaction between molecules and their environment, since every molecule will interact with its host in a slightly different manner. This is particularly the case for molecules in a glassy environment such as a protein. The mathematical model that best describes inhomogeneous broadening is a Gaussian function. The fluorescence emission spectrum of a single type of emitter at room temperature corresponds typically to a single electronic transition and therefore features as a single, broad band. A single Gaussian function provides an adequate description of such a spectrum.

It is important to take into account that energy does not scale linearly with wavelength. It is therefore not correct to fit a single-band fluorescence spectrum on the wavelength scale with a standard Gaussian function. Since energy scales linearly with wavenumber, it is more correct to convert the axes first to the wavenumber scale and then fit the spectrum with a Gaussian function. However, such conversion is not trivial, because when the measuring instrument uses a wavelength scale it bins the fluorescence counts into equal wavelength bins but when the spectrum is presented on a wavenumber scale equal wavenumber intervals are used. The fluorescence intensity therefore has to be scaled to be displayed correctly on a linear wavenumber scale. The scaling factor is λ^2. However, multiplication with λ^2 leads to selective enhancement of the long-wavelength side of the spectrum, which often shifts the peak position to the red.[24]

Diffraction of light is linear in wavelength while dispersion is linear in energy (or wavenumber). Most modern-day optical spectroscopy instruments use diffraction gratings and therefore present the absorption or emission on the wavelength scale. The rule of thumb is to use the scale of the raw data and not convert between the wavelength and wavenumber scales.

A single-band fluorescence spectrum on the wavelength scale is broader on the red side and can be fitted well using a positively skewed Gaussian function. Considering a skewness b, peak amplitude A and peak wavelength λ_0, the following expression is frequently used[26]:

$$F(\lambda) = Ae^{-\frac{\ln 2}{b^2}\left[\ln\left(1+2b\frac{\lambda-\lambda_0}{\Delta\lambda}\right)\right]^2} , \tag{3.16}$$

where the width $\Delta\lambda$ is related to the full-width at half-maximum by $\text{FWHM} = \Delta\lambda \sinh b/b$. For positive skewness, $b > 0$.

An alternative approach is to use a composite function, which describes the spectral values on one side of λ_0 by a normal Gaussian function and on the other side with a skewed Gaussian function:

$$F(\lambda) = \begin{cases} Ae^{-\frac{1}{2}\left(\frac{\lambda-\lambda_0}{\Delta\lambda}\right)^2} & \text{for } (\lambda-\lambda_0)b \leq 0 \\ Ae^{-\frac{4\ln 2}{b^2}\{\ln[1+g(\lambda)(\lambda-\lambda_0)]\}^2} & \text{for } (\lambda-\lambda_0)b > 0 \end{cases} , \tag{3.17}$$

where

$$g(\lambda) = \frac{\sqrt{\ln 2}}{\Delta\lambda}\left(1 - e^{-|b|}\right). \tag{3.18}$$

The FWHM is accordingly given by

$$\text{FWHM} = \frac{\Delta\lambda}{\sqrt{\ln 2}}e^{|b|/2} + 2\Delta\lambda\sqrt{2\ln 2}, \tag{3.19}$$

where the two terms stem from the skewed and normal Gaussian contributions, respectively.

3.7.4 Regression based on probabilistic approaches

The major drawback of the standard least-squares and maximum likelihood estimation approaches to data analysis is their dependence on a predetermined model. In these approaches, regression analysis can be performed using different mathematical functions and the one providing the best fit is typically considered the mechanistic model. But this is not necessarily the correct model. Furthermore, these approaches may even fail to predict a unique physical model in the case when different mathematical functions give rise to equivalent goodness-of-fit values (e.g. R^2 or reduced χ^2 values). Probabilistic approaches, also known as distributive approaches, provide a means around these shortcomings. They can be used for data exploration to put the data in such a form that the best correlations between the measured parameters can be identified, from which the most appropriate physical model can be derived. Such correlations can be found using analytical methods such as cluster analysis, to identify patterns or groupings in the data, or correspondence analysis, to identify possible relationships amongst specific model parameters.

Even if the best mechanistic model is known, some of the parameters in the model that are assumed to be independent may actually be correlated, especially if they describe related properties of the measured system. This phenomenon is known as multicollinearity. The strong correlation between the skewness and width of a skewed Gaussian function (Eq. (3.16)) represents one such example. A correlation between model parameters, even if it is weak, increases the variance in the estimates of the regression parameters and hence also the uncertainty in the fit. Probabilistic approaches provide a way to describe the system completely in terms of uncorrelated parameters.

Finally, many modern-day laboratory instruments give access to large numbers of variables of the investigated system. It is often impossible to display the relationship amongst all variables on a 2- or 3-dimensional plot and it is often redundant to investigate the relationship between all pairs of variables. It is therefore important to compress the information in such a way that it can be displayed in a straightforward manner while the most essential information is retained. An appropriate method is therefore required to determine which parameters should be selected to display the most important correlations in the data. Once again, probabilistic approaches are the ideal methods to do this. They are helpful to identify, for example, which features of a spectrum are the most important ones.

Pure probabilistic approaches make no assumptions about the physical model before data fitting but a physical model can be obtained using these approaches in combination with cluster or correspondence analysis. Once a plausible physical model is obtained, regression can be performed. The three most commonly used probabilistic methods are principal component analysis (PCA), the maximum entropy method (MEM) and partial least-squares regression (PLS).

Principal component analysis (PCA)

The aim of PCA is to express the data with as few parameters as possible that give as much information as possible. This method makes use of dimension reduction to reduce multi-dimensional data to lower dimensions while retaining most of the information. Loss of information during dimension reduction is minimised by identifying a set of parameters, known as the principal components, which capture the variation in the data as sufficiently as possible. The principal components are selected such that they are independent. Their independence ensures that redundancy in the information is removed and the same information can consequently be described by fewer parameters. The principles and strength of this technique are best illustrated using an example.

Consider a large set of (hypothetical) spectra from carotenoid pigments (Figure 3.8(a)). Carotenoid spectra have a characteristic three-peak structure. We are interested in finding all possible correlations between the spectra in our data set. We may start by identifying the three peaks of each spectrum. Since the positions are not always clear to find by eye, we may choose to fit each spectrum with three Gaussian functions (Figure 3.8(b)), which will provide us with three amplitudes, three peak positions, and three widths for every spectrum, i.e. nine variables. For every spectrum, the three amplitudes are then plotted as a single data point in 3-dimensional space, where the three coordinate axes represent amplitudes and are mutually orthogonal. For N spectra in the set we get N data points. We now investigate the whole distribution of data points and find the direction in our 3-dimensional space along which the data points have the broadest distribution, i.e. the highest variance (Figure 3.8(c)). This direction defines the axis of the first principal component. The axis of the second principal component is directed along the broadest distribution of the data points perpendicularly to the axis of the first principal component. The axis of the third principal component is obtained similarly and is orthogonal to the first two axes. The original coordinate system has now been transformed into a new coordinate system defined by the three principal components.

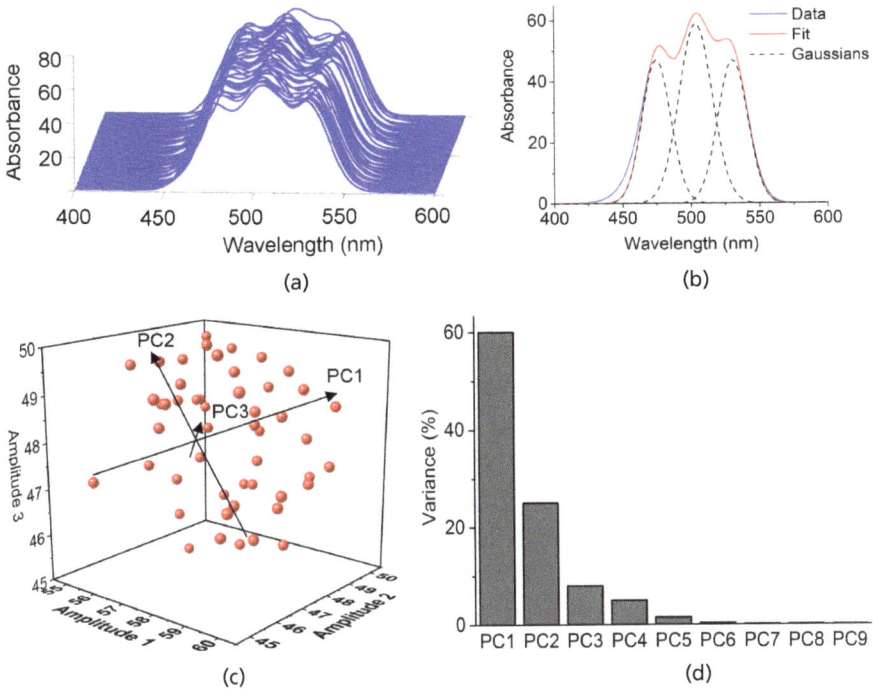

Figure 3.8: Illustration of Principal Component Analysis. (a) Data set of triple-peak spectra. (b) Deconvolution of the first spectrum using three Gaussian functions. (c) Relationship between the amplitudes of the three resolved bands of all spectra. The directions of the three principal components (PC1, PC2 and PC3) are indicated. (d) Contribution of each principal component to the total variance.

Next, we expand the description by considering the nine variables of each spectrum as a single data point in 9-dimensional space and similarly obtain the principal components. This cannot be visualised anymore, but the mathematical formulation is identical to our 3-dimensional problem, only expanded to nine dimensions. Now we only consider those principal components that correspond to a meaningful amount of variance, which are the first four components (Figure 3.8(d)). This means that the original 9-dimensional data set has been projected onto a lower-dimensional space that retains 99% of the information. The main reason why this projection is possible is redundancy in the original data. In other words, the nine variables of each spectrum are not completely independent, but some level of correlation exists. In contrast, the principal components are completely uncorrelated because of their orthogonality. Furthermore, since the principal components correspond to the highest (remaining) variance in the data, the maximum amount of information of the original data is transferred during the coordinate transformation when determining the principal components. Hence, the principal components represent the number of independent variables that describe the data.

The redundancy in the spectral variables is particularly evident when using skewed Gaussian functions to fit the spectra. Consider Eq. (3.16). There is a direct correlation between the width and skewness of a skewed Gaussian function and only one of the two variables should be included in the analysis. PCA would automatically remove such redundancy.

By first resolving the main spectral bands based on a multi-Gaussian fit introduces assumptions. For example, the spectra are assumed to consist of three spectral bands and the three bands are assumed to have a Gaussian distribution. These assumptions are reasonable approximations but are not entirely correct. Figure 3.8(b) indicates that the spectral wings of the data are broader than the fits. More generally, this approach does not make use of the full strength of PCA, which works best if no assumptions are made prior to data processing. The best approach would be to consider only the raw data, which comprises absorbance at a large number of wavelengths. This may give hundreds or even thousands of variables per spectrum. According to the PCA approach, each variable corresponds to a different dimension in the multi-dimensional space. However, most of the variables are connected, for example many of the amplitude variations are correlated, so that only a few principal components will be obtained. Use of 2–4 principal components typically retains >90% of the information content.

Principal component regression

In standard least-squares analysis, regression is performed using Eq. (3.13), considering N regression variables. In principal component regression, we wish to describe the different relationships between the experimental data as well as possible but using as few regression variables as possible. This is done by first performing PCA to identify p principal components, where $p \leq N$. Each row vector $\mathbf{x}_j$ of the matrix $\mathbf{X}$ in Eq. (3.13) is then expressed as a linear combination of the principal components contained in vector $\mathbf{z}_j$. Regression is then performed using a new model,

$$\mathbf{y} = \mathbf{Zb}, \qquad (3.20)$$

where $\mathbf{Z}$ is an $m \times p$ matrix consisting of row vectors $\mathbf{z}_j$, and $\mathbf{b}$ is a vector containing the new regression coefficients for $\mathbf{Z}$. If one wishes to investigate the original regression coefficients contained in vector $\mathbf{a}$ in Eq. (3.13), $\mathbf{b}$ can be transformed back to $\mathbf{a}$.

Maximum entropy method (MEM)

When a complex model is simplified, information is inevitably lost. The key in probabilistic data analysis approaches is to eliminate redundancy by combining correlated variables, which, in most cases, leads to a reduction in the degrees of freedom of the mathematical model. In PCA, minimising the number of principal components generally receives greater priority, whereas in MEM, larger emphasis is placed on minimising the loss of information during dimension reduction. Minimum information loss is equivalent to maximum entropy and in MEM, the Shannon–Jaynes entropy is used as a measure of the amount of information.

The goal in MEM is to transform each row vector $\mathbf{x}_j$ in Eq. (3.13) into a new, smaller row vector $\mathbf{z}_j$, such that minimum information is lost during the transformation. The transformation from x_j to z_j is expressed by the equation $z_j = Ax_j$, where A is the transformation matrix. The Shannon–Jaynes entropy contains the elements of A and serves as a constraint function to ensure that only a single solution is obtained during the transformation and that this is the optimal solution.

Partial least-squares regression (PLS)

PLS is an improvement of PCA. The drawback of PCA is that the principal components are chosen to explain the data matrix X in Eq. (3.13) but may not be the best choice for calculating y. PLS solves this problem by finding principal components that correspond to the maximum covariance of x_j and y. In other words, the directions of the principal components in PLS correspond to the direction of maximum variation in the data of both x_j and y. Maximum variation in x_j and y is not found separately but each principal component corresponds to the direction of strongest correlation between the two data sets. After identifying the principal components, regression is performed similarly as described for PCA above. PLS is a particularly useful data analysis method in chemometrics, where a large number of experimental variables are obtained and where the relationship between the variables is often not well understood.[27]

3.8 Key Points

- Accurate measurements in optical spectroscopy require avoiding or minimising reflected and scattered light within the spectrometer and in the sample. Sources of scattering are dust particles in the path of the light beam, stain and dust on the surfaces of optical components, in particular the cuvette, and turbidity of the sample itself by contamination with dust, larger colloidal aggregates or gas bubbles.
- Measurement of a dark spectrum and a reference sample, ideally using a dual-beam configuration, and checking

linearity with concentration are essential for the avoidance of errors in absorbance spectra.

- Fluorescence spectra can be obtained either by excitation at a fixed wavelength, typically the absorption maximum, and recording the full emitted spectrum (emission spectrum). Alternatively, the fluorescence is detected at a chosen fixed wavelength, while the exciting light is scanned through part or the full absorption spectrum (excitation spectrum).

- The fluorescence signal is often more than 1,000 times weaker than the exciting light and special measures should be taken to avoid errors. It is important to correct fluorescence spectra for the wavelength dependence of the spectrofluorometer. Excitation spectra in particular may be disproportionately distorted when not corrected.

- Strongly scattering samples such as powders, colloidal or biological samples in their native environment are measured in diffuse reflection. Since the amount of scattering scales nonlinearly with sample thickness and concentration the spectra are corrected using the Kubelka–Munk function.

- Time-resolved fluorescence spectroscopy requires pulsed excitation as well as detection, both at sub-nanosecond resolution. It is a sensitive probe of a fluorophore's interaction with its environment, such as temperature and pH. Energy transfer, charge transfer, quenching, rotational diffusion and solvent interaction are some of the processes that can be resolved.

- Since frequency and time are interconvertible through Fourier transformations, the fluorescence or phosphorescence emission lifetime τ can also be determined from measurements performed in the frequency domain. Phase fluorometry is done by modulating the intensity of the excitation light at a frequency comparable to the time scale of the emission. A time delay between excitation and emission is reflected in a phase shift and a reduced amplitude of e.g. a sinusoidal signal in the detection channel.

- Pump-probe spectroscopy is a two-pulse technique in which an intense first pulse excites a sizeable fraction of the chromophores. A second pulse with a variable delay then records the absorption spectrum of the sample. Any difference to the initial absorption spectrum tells about chemical and physical changes induced by the pump pulse as a function of time.
- Advanced optical spectroscopy pushes the limits of temporal and spatial resolution and sensitivity to sub-femtoseconds and to the detection of single molecules in a volume of one femtolitre.
- Use of polarised excitation and detection provides orientational information of the chromophores, the surface topology, the local viscosity of the solvent, membrane fluidity, conformational changes of proteins, configurational changes of chromophores, association reactions between biological molecules, denaturation of proteins and DNA, rotational diffusion of molecules, and energy transfer between molecules.
- Fitting residuals instead of goodness-of-fit values, such as the R^2 and χ^2 values, should be used to verify the aptness of a physical model during regression.
- Care has to be taken in quantitative analysis when the wavelength axis is transformed to frequency (wavenumber, energy) units, since the bin width is constant in the original unit but not in the transformed space, which affects the apparent intensities and skews band shapes. Furthermore, if error bars are assumed symmetric before linearisation of a function they no longer are after this transformation. This complicates least-squares analysis and error propagation.
- Furthermore, if error bars are assumed symmetric before linearisation of a function they no longer are after this transformation. This complicates least-squares analysis and error propagation.
- The main weakness of standard least-squares analysis is its dependence on a predetermined physical model.

This problem is avoided in probabilistic data analysis approaches, which provide tools to find the most significant correlations between experimental parameters and to avoid redundancy in describing the experimental data.

General Reading

- J. R. Lakowicz, *Principles of Fluorescence Spectroscopy*, 3rd Ed., Springer, Singapore, 2006.
- B. Valeur, *Molecular Fluorescence: Principles and Applications*, Wiley-VCH Verlag GmbH, 2001, ISBN: 9783527299195 (Electronic), 9783527600243 (Hardcover).
- N. V. Tkachenko, *Optical Spectroscopy: Methods and Instrumentations*, Elsevier Science, Amsterdam, 2006.
- W. Becker, *Advanced Time-Correlated Single Photon Counting Techniques*, Springer, 2005.
- D. V. O. O'Connor, D. Phillips, *Time-correlated Single Photon Counting*, Academic Press, London, 1984.
- M. Y. Berezin, S. Achilefu, Fluorescence lifetime measurements and biological imaging, *Chem. Rev.*, 2010, 110, 2641–2684.
- J. A. Ross, D. M. Jameson, Time-resolved methods in biophysics. 8. Frequency domain fluorometry: Applications to intrinsic protein fluorescence, *Photochem. Photobiol. Sci.*, 2008, 7, 1301–1312.
- A. H. Zewail, Femtochemistry: Atomic scale dynamics of the chemical bond using ultrafast lasers (Nobel Lecture), *Angew. Chem. Int. Ed.*, 2000, 39, 2586–2631.
- B. Valeur, J. C. Brochon, *New Trends in Fluorescence Spectroscopy: Applications to Chemical and Life Science*. Springer, New York, 2001.
- A. Stortfelder, J. B. Buijs, J. Bulthuis, C. Gooijer, G. Van der Zwan, Fast-gated intensified charge-coupled device camera to record time-resolved fluorescence spectra of tryptophan, *Appl. Spectr.*, 2004, 58, 705–710.
- W. W. M. Wendlandt, H.G. Hecht, *Reflectance Spectroscopy*, Interscience Publishers/John Wiley, NY, 1966.
- G. Kortüm, *Reflectance Spectroscopy / Reflexionsspektroskopie*, Springer, Berlin, 1969.
- D. J. Dahm, K. D. Dahm, *Interpreting Diffuse Reflectance and Transmittance*, IM Publications, Chichester, 2007.

- J. A. N. T. Soares, Introduction to optical characterization of materials. In: *Practical Materials Characterization,* M. Sardela, (ed.), Springer Science+Business Media New York 2014, DOI 10.1007/978-1-4614-9281-8_2.
- J. J. Snellenburg, S. P. Laptenok, R. Seger, K. M. Mullen, I. H. M. van Stokkum, Glotaran: A Java-based graphical user interface for the R package TIMP. *J. Stat. Softw.,* 2012, 49, 1–22. URL: http://www. jstatsoft.org/v49/i03/.
- C. Slavov, H. Hartmann, J. Wachtveitl, Implementation and evaluation of data analysis strategies for time-resolved optical spectroscopy. *Analyt. Chem.,* 2015, 87, 2328–2336.
- J. Beechem, E. Gratton, M. Ameloot, J. Knutson, L. Brand, The global analysis of fluorescence intensity and anisotropy decay data: Second generation theory and programs. In: *Topics in Fluorescence Spectroscopy,* J. R. Lakowicz, (ed.), 1991, 1, 241–305.
- A. R. Holzwarth, Data analysis of time-resolved measurements. In: *Biophysical Techniques in Photosynthesis. Advances in Photosynthesis and Respiration,* J. Amesz, A. J. Hoff, (eds.)., 1996, 3, 75–92.
- I. H. M. van Stokkum, D. S. Larsen, R. van Grondelle, Global and target analysis of time-resolved spectra. *BBA — Bioenergetics,* 2004, 1657, 82–104.
- E. Gratton, D. M. Jameson, R. D. Hall. Multifrequency phase and modulation fluorometry. *Ann. Rev. Biophys. Bioeng.,* 1984, 13, 105–124.
- Glotaran: http://glotaran.org/index.html or http://glotaran.org/wiki. html.
- Optimus: https://optimusfit.org/index.php/data-analysis/global-target-analysis.

References

1. http://www.hamamatsu.com/, accessed 30 Dec. 2017.
2. https://www.hamamatsu.com/resources/pdf/etd/LIGHT_SOURCE_ TLSZ0001E.pdf.
3. M. van de Ven, M. Ameloot, B. Valeur, N. Boens, *J. Fluorescence,* 2005, 15, 337–413.

4. B. Valeur, *Molecular Fluorescence: Principles and Applications*, Wiley-VCH, Weinheim, 2001, ISBN: 9783527299195 (Electronic), 9783527600243 (Hardcover).
5. A. V. Fonin, A. I. Sulatskaya, I. M. Kuznetsova, K. K. Turoverov, *PLOS One*, 2014, 9, e103878. https://doi.org/10.1371/journal.pone.0103878.
6. F. C. Jentoft, *Adv. Catal.*, 2009, 2, 129–211.
7. B. M. Weckhuysen, R. A. Schoonheydt, *Catal. Today*, 1999, 49, 441–451.
8. P. Kubelka, F. Munk, *Z. Tech. Phys.*, 1931, 12, 593–601.
9. G. Kortüm, *Reflectance Spectroscopy / Reflexionsspektroskopie*, Springer, Berlin 1969.
10. G. Kortüm, H. Schottler, *Z. Elektrochem.*, 1953, 57, 353–361.
11. E. Ghadiri, S. M. Zakeeruddin, A. Hagfeldt, M. Grätzel, J.-E. Moser, *Sci. Rep.*, 2016, 6, 24465.
12. J. Torrent, V. Barrón, Diffuse reflectance spectroscopy, Chapter 13. In: *Methods of Soil Analysis,* Part 5. Mineralogical Methods. A. L. Ulery, L. R. Drees, Eds., Soil Science Society of America Book Series, Madison, USA, 2008.
13. R. W. Frei, *J. Res. National Bureau of Standards — A. Physics and Chemistry*, 1976, 80A, 551–565.
14. P. Schwach, N. Hamilton, M. Eichelbaum, L. Thum, T. Lunkenbein, R. Schlögl, A. Trunschke, *J. Catal.*, 2015, 329, 574–587.
15. K. Amakawa, L. Sun, C. Guo, M. Hävecker, P. Kube, I. E. Wachs, S. Lwin, A. I. Frenkel, A. Patolla, K. Hermann, R. Schlögl. A. Trunschke, *Angew. Chem. Int. Ed.*, 2013, 52, 13553–13557.
16. D. Maganas, A. Trunschke, R. Schlögl, F. Neese, *Faraday Discuss.*, 2016, 188, 181–197.
17. D. G. Barton, M. Shtein, R. D. Wilson, S .L. Soled, E. Iglesia, *J. Phys. Chem. B,* 1999, 103, 630–640.
18. A. Escobedo Morales, E. Sanchez Mora, U. Pal, *Revista Mexicana de Fisica*, 2007, 53, 18–22.
19. A. Stortelder, J. B. Buijs, J. Bulthuis, C. Gooijer G. van der Zwan, *Appl. Spectr.*, 2004, 58, 705–710.
20. https://www.hamamatsu.com/resources/pdf/sys/SHSS0006E_STREAK.pdf.
21. M. Szabelski, R. Luchowski, Z. Gryczynski, P. Kapusta, U. Ortmann, I. Gryczynski, *Chem. Phys. Lett.*, 2009, 471, 153–159.

22. R. Luchowski, M. Szabelski, P. Sarkar, E. Apicella, K. Midde, S. Raut, J. Boreido, Z. Gryczynski, I. Gryczynski, *Appl. Spectrosc.*, 2010, 64, 918–922.

23. B. van Oort, A. Amunts, J. W. Borst, A. van Hoek, N. Nelson, H. van Amerongen, R. Croce, *Biophys. J.*, 2008, 95, 5851–5861.

24. J. Lakowicz, *Principles of Fluorescence Spectroscopy*, 2nd ed., Kluwer, New York, 1999.

25. D. A. Smith, G. McKenzie, A. C. Jones, T. A. Smith, *Methods Appl. Fluoresc.*, 2017, 5, 042001.

26. R. D. B. Fraser, E. Suzuki, *Anal. Chem.*, 1969, 41, 37–39.

27. D. G. Kleinbaum, L. L. Kupper, K. E. Muller, A. Nizam, *Applied Regression Analysis and Multivariable Methods*, 3rd ed., Duxbury Press, Pacific Grove, CA.: 1998.

Chapter 4

Principles of Optical Spectroscopy Demonstrated for a Set of Rigid Merocyanine Dyes

4.1 Concept

Here we discuss a systematic set of organic merocyanine dyes (Figure 4.1) which are rigid because of the *all-trans* conjugated π-system that extends between the nitrogen atom of the piperidine donor unit and the variable acceptor unit. While the piperidine donor end is identical for all molecules, there are three different acceptor groups, and the conjugated π-system comes in three different lengths. We discuss the influence of these structural variations on the electronic and the spectroscopic properties such as absorption and fluorescence wavelengths and on the magnitude of the extinction coefficient, and we explore the effect of the solvent environment on these properties.

The entire chapter is based on the PhD thesis of Katharina Kress.[1]

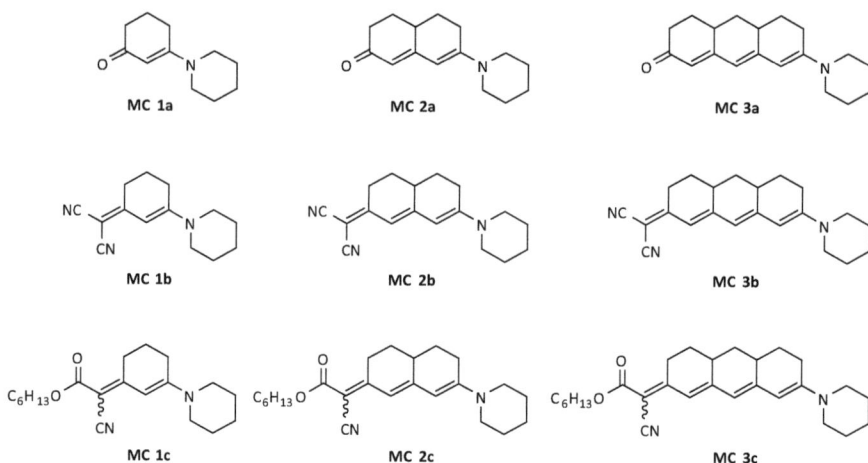

Figure 4.1: Set of three groups of rigid merocynanine (**MC**) dyes, all with the same piperidine donor unit, but with different acceptor units (keto group **a**, malodinitrile **b**, and cyanoacetate with C_6H_{13} end group **c**). Each group comes in three different sizes of the chromophore, namely with one (**1**), two (**2**), and three (**3**) cyclohexene ring units which conjugate into the acceptor unit and the nitrogen of the piperidine donor unit.

4.2 Absorption Properties

The lower limit of detected wavelengths of the human eye is near 400 nm. This is verified by comparison of Figures 4.2 and 4.3. The solutions of **MC 1a** and **MC 2a** are colourless, and the absorption is almost entirely below 400 nm. **MC 1b** and **MC 1c** have a faint colour because the absorption edge is just above 400 nm. **MC 2b** and **MC 2c** are intensely yellow since their absorption bands peak around 500 nm and have a long tail towards the blue. This means that they remove mostly the blue part of the electromagnetic spectrum. The combination of the remaining colours (mainly green, yellow and red) appears yellow. **MC 3a** appears clear yellow, because its broad absorption band is shifted to shorter wavelengths and is less intense. The solution therefore absorbs a larger fraction of blue light than **MC 2b** and **MC 2c**. The absorption of **MC 3b** and **3c** is

Figure 4.2: Solutions of the nine merocyanine dyes of Figure 4.1 (**MC 1abc, MC 2abc, MC 3abc**, from left to right) in DMSO at a concentration c $\approx 10^{-5}$ mol L^{-1}.

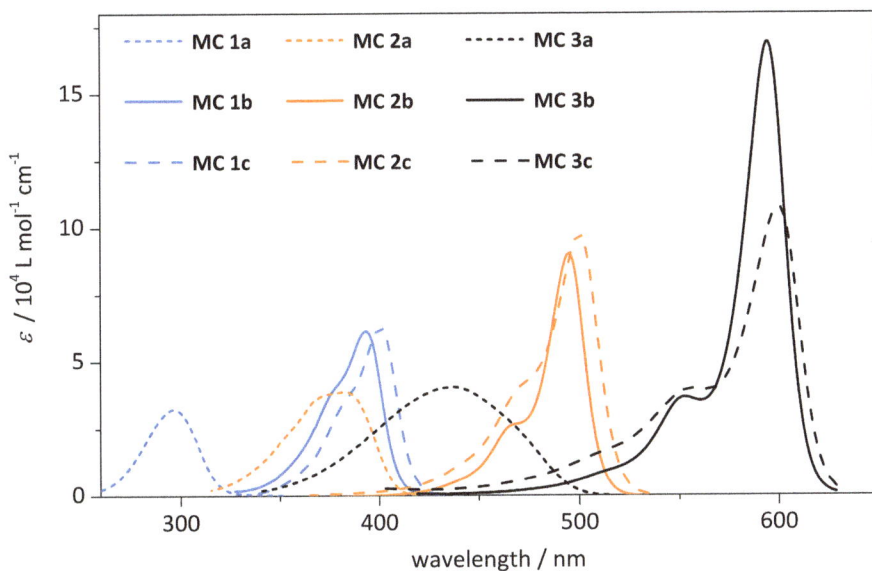

Figure 4.3: UV–Vis absorption spectra of the set of nine merocyanine dyes in DMSO displayed in Figure 4.2 at a concentration of 10^{-5} mol L^{-1} (reprinted with permission from Ref. [2]. © (2014) John Wiley and Sons).

very intense around 600 nm, with a long tail towards the blue. The spectra filter out yellow, a large amount of green and some blue light, which leads to an intense blood red colour for **MC 3b** and wine red for **MC 3c**. The unabsorbed blue light is responsible for darkening of the strong red colour.

It is obvious from the spectra that the wavelength of the absorption maximum depends on the size of the delocalised system, shifting to higher values by roughly 100 nm for each additional double bond in the conjugated system. This is under-stood based on the simple quantum mechanical model of the particle in a 1-dimensional box, where the energy spacing diminishes when the length of the box increases.

It is furthermore obvious from Figure 4.3 that the extinction coefficients at the band maxima increase with the size of the chromophore. This is expected since the extinction coefficient scales linearly with the absorption cross-section (Eq. (2.7)), which increases with the geometrical cross-section within a family of similar chromophores (see Section 2.3). The corresponding val-ues in Table 4.1 reveal that this dependence is qualitatively

Table 4.1: Experimental extinction coefficients ε of merocyanine dyes at band maximum and oscillator strength f derived from integrated band area in DMSO.

dye	ε_{max} 10^4 L mol^{-1} cm^{-1}	f
MC 1a	3.2	0.54
MC 2a	3.9	0.62
MC 3a	4.1	0.60
MC 1b	6.1	0.75
MC 2b	9.0	0.80
MC 3b	16.9	0.86
MC 1c	6.3	0.77
MC 2c	9.7	0.82
MC 3c	10.8	0.84

correct but not quantitatively rigid. The fact that the oscillator strength f is close to unity confirms that the observed transition is allowed. Strictly forbidden transitions have oscillator strengths typically in the order of 10^{-5}.

4.3 Solvatochromic Absorption Properties

The normalised absorption spectra of the **MC 3a** and **MC 3c** dyes, shown for five different solvents in Figure 4.4, reveal enormous shifts of the absorption bands. The **MC 1** and **MC 2** dyes display analogous behaviour with a less pronounced shift. Toluene is the least polar amongst the chosen solvents, and the absorption maxima are observed at the shortest wavelengths. With increasing solvent polarity, the bands shift to

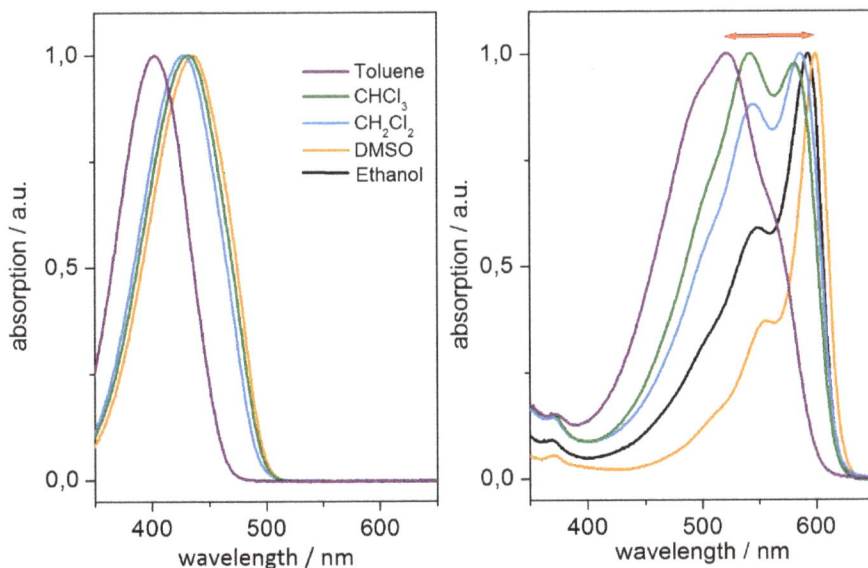

Figure 4.4: Normalised absorption spectra of dye **MC 3a** (left) and **MC 3c** (right) in different solvents at room temperature. The malonitrile derivatives **MC 3b** behave much like **MC 3c** and are therefore not shown. The red arrow indicates the splitting into vibrational sub-bands.

progressively longer wavelengths. Based on the explanation given in Figure 2.11(a) this represents a bathochromic shift and means that the excited state S_1 is more polar than the S_0 ground state. The shift from 400 to 450 nm for **MC 3a** represents an additional stabilisation due to solvation of 34 kJ mol^{-1}, which is an impressive amount, while the band shape undergoes little change. In contrast, for **MC 3c**, the 0-0 transition remains almost unshifted, but the second component, which is a 1←0 vibrational sub-band, shows enormous changes in its intensity relative to that of the 0-0 component.

The vibrational splitting by 50 nm for **MC 3c** corresponds to a vibrational quantum of *ca.* 1500 cm^{-1}, which is in the range of aromatic and conjugated polyene C=C bonds. As we will see below, excitation of a $\pi^*\leftarrow\pi$ transition weakens these delocalised bonds, which affects the carbon nuclear coordinates (q_{01} in Figure 2.10) and therefore the Franck–Condon factor (Eq. (2.14)). The strength of other bonds remains almost unaffected, which renders their Franck–Condon factor zero. Therefore, it is often a single mode that shows up as vibrational sub-structure in optical spectra of organic dyes, and it often relates to a delocalised C=C bond of the chromophore. It is obvious from Figure 2.10 that the intensity of the vibrational sub-transition is sensitively affected by the Franck–Condon factor.

4.4 Electronic Structure

Electronic excitation changes the electronic structure of a molecule. This leaves its footprint in the spectra, but the information is difficult to discern, and important support for the interpretation comes from quantum chemical calculations. Here we report the results of density functional theory (DFT) calculation.[2] The **MC 3b** molecule will serve for an exemplary discussion.

Figure 4.5 displays the basic structure in the ground state and its changes on excitation to S_1. The conjugated system

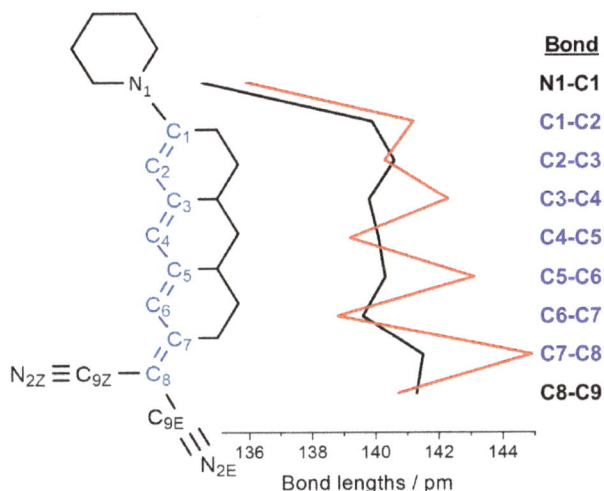

Figure 4.5: Calculated bond length alterations of dye **MC 3b** on excitation from the S_0 ground state (black) to the S_1 excited state (red), revealing that the nominal C=C double bonds increase in length on excitation while the nominal single bonds remain nearly unchanged. The C_8–C_9 bond lengths are nearly the same for the E and the Z nitrile groups.

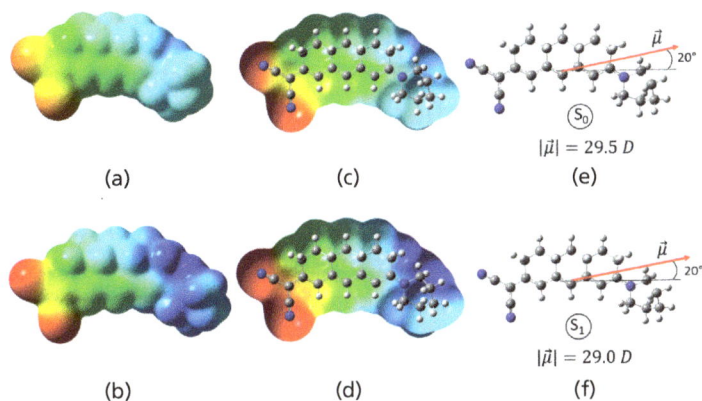

Figure 4.6: DFT calculations for the dye molecule **MC 3b**. Molecular electrostatic potential in toluene (a) and in DMSO (b) with colour scale ranging from −0.066 C (red) to +0.066 C (blue). Molecular electrostatic potential in the S_0 ground state (c) and the S_1 excited state (d), with colour scale from −0.12 C (red) to +0.12 C (blue). Electric dipole moment in the ground state (29.5 D,e) and in the S_1 state (29.0 D,f).

shows C=C bond lengths of *ca.* 140 pm, with little variation and similar to the values in benzene. The single and double bonds in the drawing could be swapped, but this would lead to a zwitterionic structure. The longest wavelength transition is of $\pi^* \leftarrow \pi$ character, meaning that an electron from the HOMO π bond is promoted into the LUMO π^* antibonding orbital. This weakens the π system, which is seen in considerable elongations of the nominal double bonds and an elongation of the entire system from C_1 to C_9 by 7 pm. This offset in the coordinates in the S_1 *vs.* the S_0 state enters the Franck–Condon expression (Eq. (2.14)) and leads to the presence of vibrational substructure with a frequency characteristic for delocalised C=C systems.

Figure 4.6 displays the calculated molecular electrostatic potential, which provides a measure of the molecular charge polarisation, showing impressively the negative end at the nitrile acceptor moieties. The electrical dipole moment increases dramatically from vacuum ($\varepsilon = 0.0$, $\mu = 17.7$ D) over toluene solvent ($\varepsilon = 2.3$, $\mu = 22.5$ D) to DMSO ($\varepsilon = 48.0$, $\mu = 29.5$ D), demonstrating notably how the solvent influences electron distribution in a delocalised system. Interestingly, the dipole moment in DMSO is nearly unaffected by the $\pi^* \leftarrow \pi$ electronic excitation, reflecting the fact that the redistribution of electron density occurs mainly between neighbouring C=C bonds.

The energy of four levels on either side of the HOMO–LUMO gap, and LCAO–MO representation of the HOMO, the LUMO and the LUMO+2 molecular orbitals are displayed in Figure 4.7. The contributing atomic orbitals of the dominant electric dipole allowed HOMO–LUMO transition are of π symmetry and located along the conjugated chain. The nodes of the HOMO perpendicular to the molecular plane are near C_7, C_5, C_3, and C_1, while those of the LUMO are near C_8, C_6, C_4, and C_2. The LUMO+2 orbitals have no node in the molecular plane, and they are localised mostly near the dicyano acceptor group. The LUMO+2 ← HOMO transition is of $n \leftarrow \pi$ type and therefore

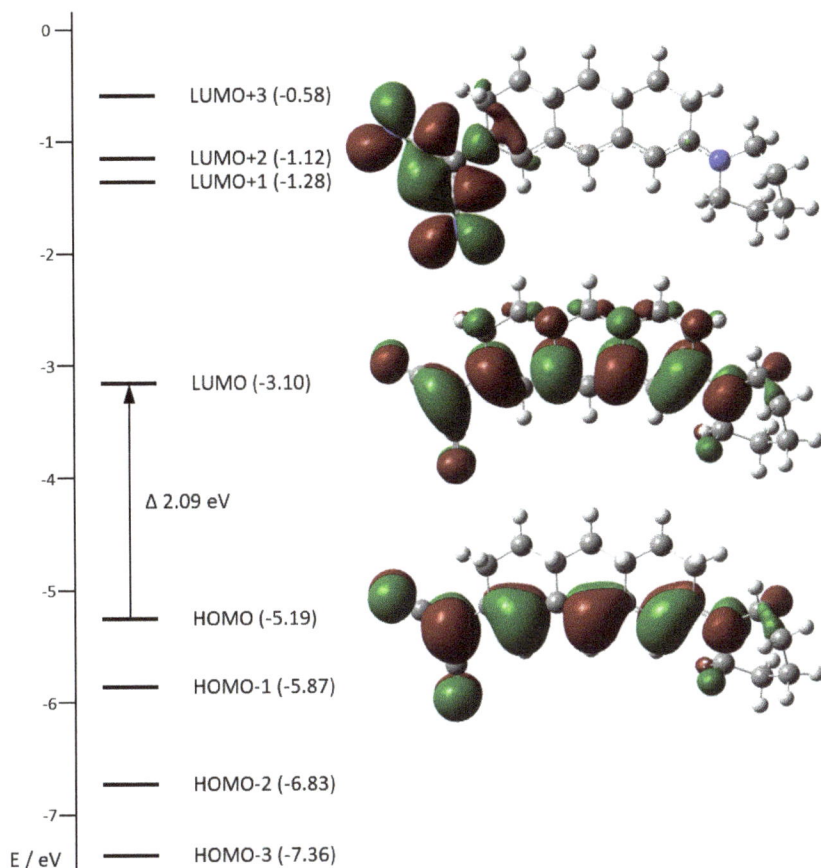

Figure 4.7: DFT calculation of **MC 3b** with energy levels and LCAO–MO image of the HOMO, the LUMO, and the LUMO+2 molecular orbitals.

forbidden, which is confirmed by the very small calculated oscillator strength, $f = 0.0003$.

4.5 Fluorescence Spectra

In the same sense as absorption spectra, also fluorescence spectra exhibit solvatochromism.[3] Emission occurs from the vibrational ground state of the excited S_1 state. For simple cases, one expects mirror symmetry according to the Franck–Condon principle described in Figure 2.10. However, a liquid solvent relaxes by reorientation of its dipole moments to adapt to the new electron distribution in the dye following an electronic transition, thereby minimising the energy of the system (as described in Figure 2.11). Depending on the solvent polarity and dynamics, this relaxation may take place at the same time scale as fluorescence emission of organic dyes. In these cases, emission spectra display a time-dependent Stokes shift,[4] reflecting the S_1 state in the unrelaxed solvent at early times and the relaxed state at later time. Relaxation affects the nuclear coordinates of the dye (q_{01} in Figure 2.10) and therefore the intensity distributions between the vibrational sub-bands. The vibrational sub-bands become solvent-dependent as shown impressively in Figure 4.8, where the minor transition to the $v'' = 1$ state (compare Figure 2.11) near 650 nm becomes increasingly important as one changes from the polar DMSO (relative dielectric constant $\varepsilon_r = 48.0$) to EtOH ($\varepsilon_r = 34.3$), $CHCl_3$ ($\varepsilon_r = 9.1$), and the nearly non-polar toluene ($\varepsilon_r = 2.3$). These effects can be elucidated in complex time-dependent measurements[4] and simulated in demanding quantum-chemical calculations.[5–7] The time-dependence of fluorescence due to solvent relaxation therefore provides an explanation for the often-observed tail towards longer wavelengths in emission spectra.

Figure 4.9 compares absorption and emission spectra of two dyes in the polar solvent DMSO. **MC 3a** displays a relatively large Stokes shift of 59 nm although its calculated ground state dipole moment amounts to a modest value of 17.5 D, significantly less than the 29.5 D of **MC 3b** that shows a Stokes shift of only 13 nm. The explanation lies in the near negligible change of the dipole moment of **MC 3b** on excitation (see Figure 4.6), which has the consequence that the solvent shell needs very little adaption after excitation. This obviously affects also the line widths. The solvent shell of **MC 3b** is in equilibrium and matches both the ground and the excited states. This is not the case for **MC 3a** directly after excitation, and obviously the

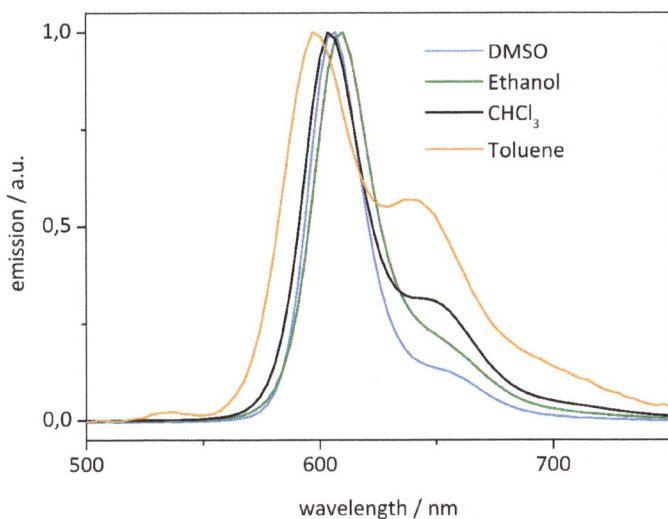

Figure 4.8: Normalised experimental bands of fluorescence emission of **MC 3b** in various solvents (excitation at the absorption maximum, e.g. at 594 nm in DMSO).

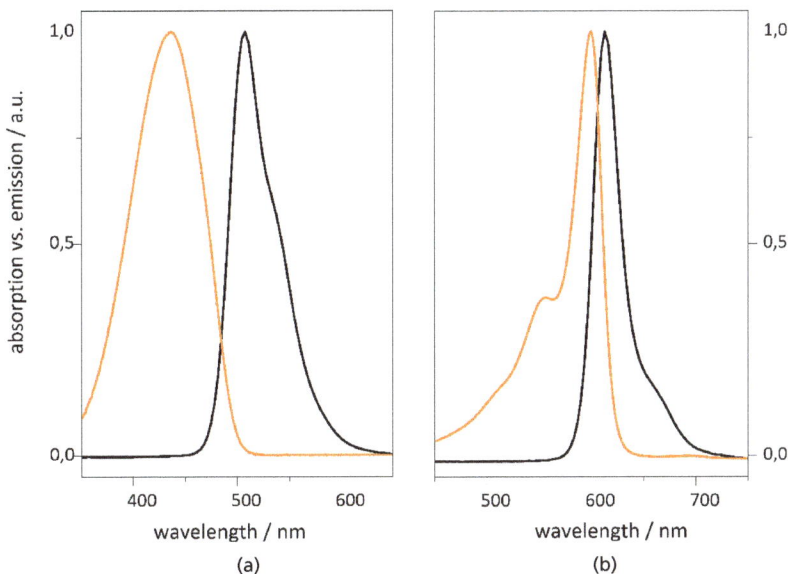

Figure 4.9: Absorption (black) and fluorescence emission spectra (orange) of the dyes **MC 3a** (a) and **MC 3b** (b) in DMSO.

details of non-equilibrium are not the same for all molecules, giving rise to slight variations of the absorption energy.

4.6 Fluorescence Quantum Yield

The process of fluorescence occurs in competition with non-radiative deactivation (internal conversion, IC) and intersystem crossing (ISC) to the triplet state. The present family of dyes shows no phosphorescence. It is plausible that ISC remains a strictly forbidden process since there are no heavy element substituents or heavy atoms in the solvent that could promote ISC by spin–orbit coupling. Thus, fluorescence probably has to compete with IC only. Figure 4.10 displays its quantum yields for the three-ring dyes in various solvents. There is an obvious dependence on the acceptor group and on the polarity of the solvent. Small and rigid acceptor groups suppress IC and enhance the fluorescence quantum yield. Polar solvents enhance this trend

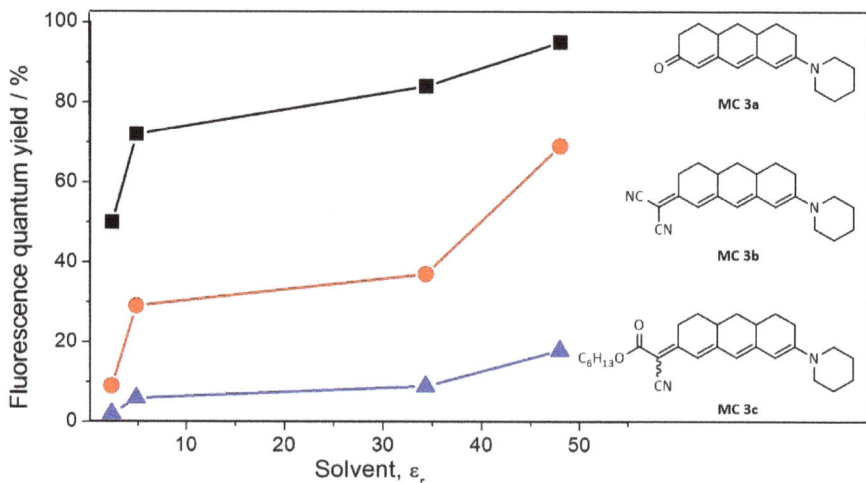

Figure 4.10: Fluorescence quantum yield of the three-ring merocyanine dyes **MC 3abc** in toluene, trichloromethane, ethanol, and DMSO.

further so that the ketone **MC 3a** reaches a 95% fluorescence quantum yield. The single- and double-ring dyes **MC 1** and **MC 2** show values ≤2%.

4.7 Orientation-dependent Absorption

So far, we have discussed the optical absorption and emission behaviour in liquid solution. We now turn our attention to dyes with preferential alignment in a liquid crystal matrix. The transition dipole moment for the longest wavelength absorption of this family of dyes is oriented along the long axis of the molecule, i.e. parallel to the delocalised π-bonds. Thus, if the molecules are aligned and the exciting light is plane-polarised parallel to the transition moment the probability of absorption is maximum. In contrast, when the plane of polarisation of the

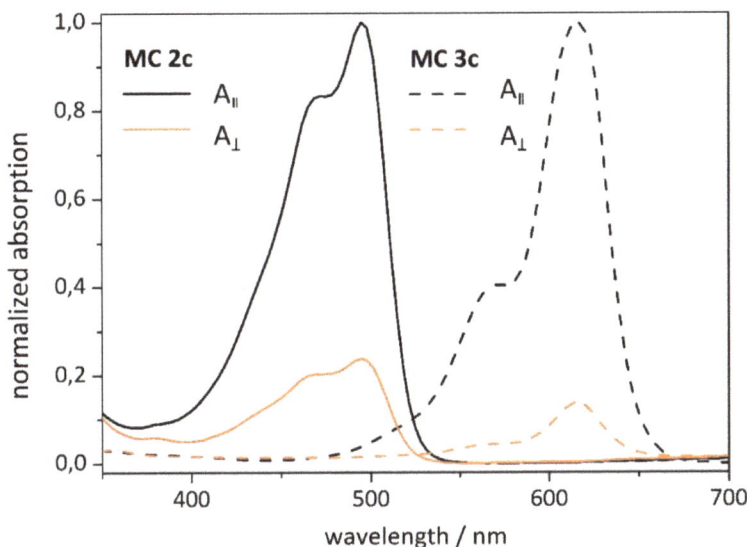

Figure 4.11: Normalised absorption of the dyes **MC 2c** and **MC 3c** in a liquid-crystalline matrix M677[1] using plane-polarised light with the electric field vector parallel (black) and perpendicular (orange) to the direction of the liquid crystal phase.

exciting light is rotated by 90° to a direction perpendicular to the transition dipole moment, the absorption should be expected to be zero for perfect alignment. However, alignment in liquid crystals is not perfect, and absorption will not normally be suppressed completely. This behaviour is verified in Figure 4.11. The ratio of absorption with light polarised parallel *vs.* absorption with perpendicularly polarised light is called the dichroic ratio, DR(λ_{max}) = $A_{\parallel}/A_{\perp}$. It amounts to 5.1 for **MC 2c** and to 7.9 for **MC 3c**, reflecting the fact that a longer molecule tends to be more strongly oriented by the matrix.

4.8 Key Points

- The longest wavelength electronic transition in organic molecules is from the HOMO (or a non-bonding orbital) to the LUMO. For saturated molecules this transition lies in the UV, but for planar unsaturated and, therefore, delocalised chromophores (i.e. the colour-determining group of a dye) the absorption is mostly in or neighbouring the visible range of the spectrum.
- The transition involves a redistribution of electron density in the plane of the molecule, and the transition dipole moment lies along the direction of this redistribution.
- Extending the size of the chromophore shifts the longest wavelength absorption band to longer wavelengths and increases the extinction coefficient.
- Solvents stabilise or destabilise the ground and excited states of molecules and because of their interaction with polar structures. This effect cause shifts in the positions of absorption and emission bands.

References

1. K. C. Kress, *Synthese und Eigenschaften rigidisierter Merocyanin-Farbstoffe: Vom quantenmechanischen Design zur Anwendung.* Verlag Dr. Hut, München, 2015, ISBN: 978-3-8439-2274-6.

2. K. C. Kress, T. Fischer, J. Stumpe, W. Frey, M. Raith, O. Beiraghi, S. H. Eichhorn, S. Tussetschläger, S. Laschat, *ChemplusChem*, 2014, 79, 223–232.
3. A. A. Ishchenko, A. V. Kulinich, S. L. Bondarev, V. N. Knyukshto, *Opt. Spectrosc.*, 2008, 104, 57–68.
4. M. L. Horng, J. A. Gardecki, A. Papazyan, M. Maroncelli, *J. Phys. Chem.*, 1995, 99, 17311–17337.
5. C-P. Hsu, Y. Georgievskii, R. A. Marcus, *J. Phys. Chem. A*, 1998, 102, 2658–2666.
6. R. Improta, V. Barone, F. Santoro, *Angew. Chem. Int. Ed.*, 2007, 46, 405–408.
7. Y. Kawashia, S. Yamamoto, T. Sakata, H. Nakano, K. Nishiyama, R. Akiyama, *J. Phys. Soc. Jpn.*, 2012, 81, SA024.

Chapter 5

Absorption and Luminescence of Semiconductor Quantum Dots

5.1 Introduction

Quantum dots (QDs) are particles that are only several nanometres in diameter (i.e. smaller than the exciton Bohr radius) (Figure 5.1), which imparts optical and electronic properties that differ from the bulk material.[1] These colloidal heteronanocrystals are typically made from semiconductor materials, such as CdSe and graphene oxide, but they may also be prepared from the highly conducting graphene. QDs were first discovered by Alexey Ekimov in 1981,[2] whilst the term "quantum dot" was coined by Mark Reed in 1986.[3]

Semiconductor QDs are typically made from elements from group III and V (indium phosphate (InP), indium arsenate (InAs), gallium arsenate (GaAs) or gallium nitride (GaN)) or from elements from groups II and VI such as zinc sulfide (ZnS), zinc selenide (ZnSe), cadmium selenide (CdSe), and cadmium telluride (CdTe).[5]

QDs have unique chemical and physical properties including size-dependent fluorescence, high fluorescence quantum yields, narrow spectral bands, independence of emission on the excitation wavelength, and stability against photobleaching.[6] Advantages of QDs over larger particles may also be linked to their large surface area to volume ratios, which provide colloidal stability, coupled to their surface activity. The QDs are small enough to exhibit quantum mechanical properties due

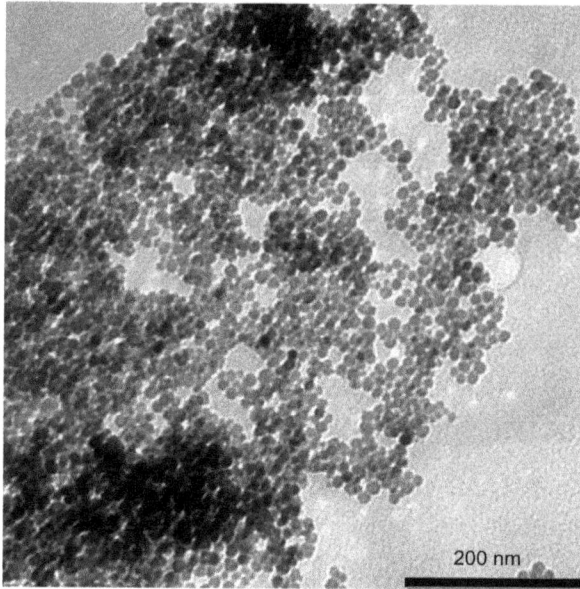

Figure 5.1: TEM image of hydrophobic CdSeTeS/ZnS QDs (reprinted with permission from Ref. [4]. © (2015) Elsevier).

to confinement of the excitons in all three spatial dimensions, reminding of the simple quantum mechanical model of a particle in a box. Due to this quantum confinement, QDs have size-dependent optical properties, which means their absorption and emission properties can be tuned by changing the particle size, shape and surface structure,[7] which is discussed more in the following sections.

QDs have thus found application in numerous fields and reviews have been published on their application in solar cells,[8] as biological labels,[9] in microelectronics,[10] in electrochemistry,[11] as well as in fluorescence sensing applications, such as the environmental detection of pesticides.[12] The latter application is elaborated on in Section 7.3.

5.2 Synthesis of Semiconductor Quantum Dots

QDs are grown in solutions via a range of synthetic routes which are briefly discussed here. The resulting material consists of inorganic nanoparticles which are overcoated with a layer of organic ligand molecules.

5.2.1 *Core quantum dots*

QDs may be prepared via the organometallic synthesis route in high temperature organic solvents, where the reaction mixture is comprised of three components: precursors, organic surfactants and solvents (Figure 5.2). The size of the resulting QDs can be varied by altering the amount of precursors and growth time. It is important to choose appropriate precursor molecules to serve as organometallic reagents (such as CdO), as they are required to decompose and produce reactive atomic or molecular species (monomers) for nanocrystal nucleation and subsequent growth. Synthesis is conducted in an inert atmosphere (under argon) in order to prevent oxidation or degradation of reagents. The organic solvent employed interacts with the precursors via Van der Waals forces and they must be able to withstand temperatures of 200–400°C. Trioctylphosphine oxide (TOPO) and octadecene are thus commonly used in QD synthesis.

The organometallic synthesis route for QDs is the most widely employed method, as the QDs thus produced have a high degree of crystallinity and a narrow size distribution, as particle growth can be well controlled by varying reaction time and temperature. In addition they have excellent luminescence properties including high photoluminescence (PL) quantum yield.

An alternative to the organometallic route to synthesise QDs (such as CdSe QDs) is direct aqueous synthesis, which avoids the use of hot organic solvents. In this approach a metal salt and a thiol compound are dissolved in water and are heated to allow for particle growth to commence. The thiol groups serve as stabilisers, as they serve as surface ligands. Although this means of synthesis is more cost effective and environmentally friendly compared to the organometallic route, it has been found that the thiol groups are typically weakly bound to the

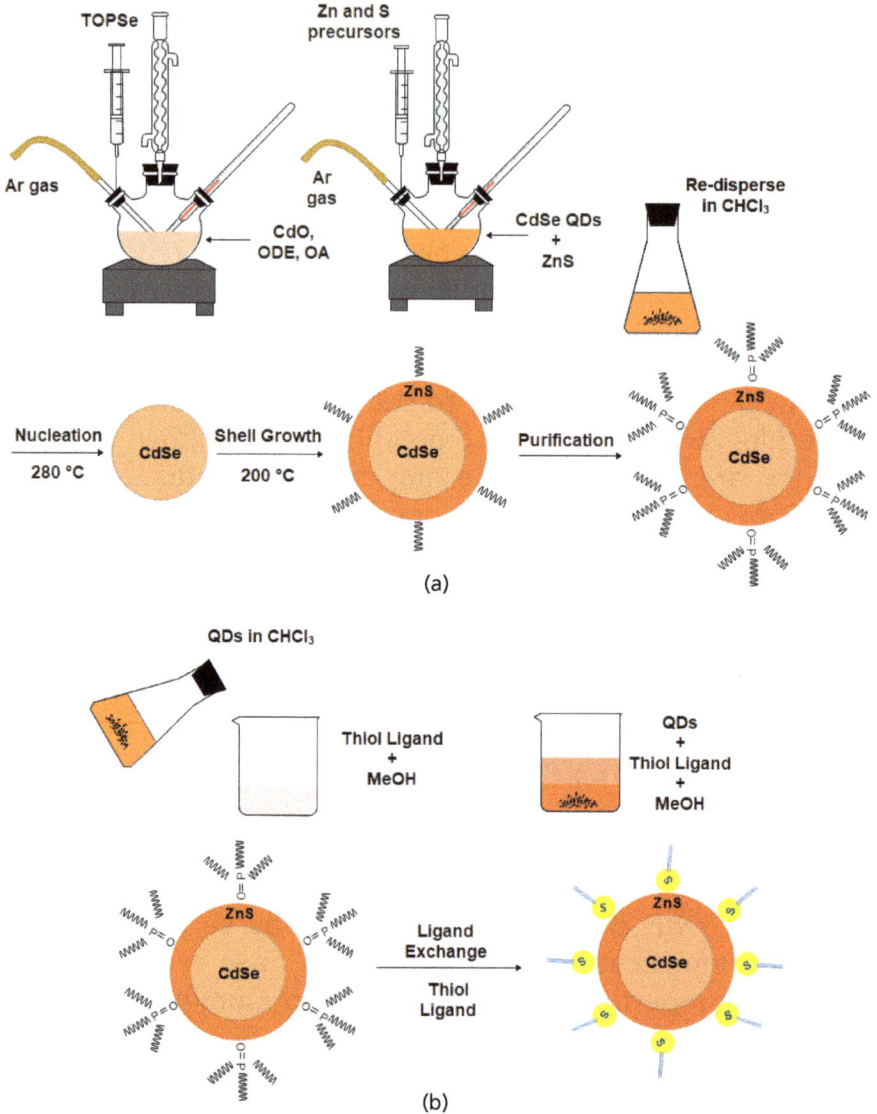

Figure 5.2: (a) Schematic of the synthesis of QDs using the organometallic approach and (b) surface modification of synthesised by thiol ligands (courtesy: Hanieh Montaseri, University of Pretoria).

QDs prepared in this manner, which results in aggregation of the QDs over time and thus a degradation in the optical properties thereof including photostability.[13]

Due to environmental considerations, the use of microorganisms to produce QDs has also been explored. This biomanufacturing or biosynthesis approach is not easily controlled, however, and thus has the disadvantage of producing a range of QD sizes within a batch, which negatively impacts the optical properties and thereby the applications thereof.

5.2.2 *Core/shell quantum dots*

In core/shell QDs, the inner core is covered with an outer shell layer of another semi-conductor material chosen with a higher band gap which enhances the optical properties of the resulting heterogeneous composite material. In addition, the overcoating of the core QD with a shell layer may reduce the leachability of the core, thereby preserving its functionality in use as well as reducing the potentially negative environmental or human health impacts associated with leaching of toxic core material such as with cadmium based QDs. In addition, the shell layer may increase the functionality, stability or dispersibility of the resulting material and reduce the surface defects thereof.

The consequent increase in QD particle size which accompanies overcoating of a core QD with a shell layer results in the fluorescence emission of the material shifting to longer wavelengths, which therefore also provides tunability in this regard (see Section 5.3.1). PL quantum yields and luminescence intensities may also be improved. Semiconductor QDs have broad absorbance peaks, therefore they can be excited using a range of wavelengths. Upon capping of a core QD with a shell layer, the absorbance peak may broaden (Figure 5.3).

It is also possible to synthesise QDs which have multiple shell layers, such as in the case of CdSeTe/CdS/ZnSe/ZnS QDs shown in Figure 5.3. The nomenclature of these materials starts with the core and separates successive layers by a slash (or by a hyphen). A shell with two cations or two anions such as CdSeTe represents an alloyed material.

For the shell to grow, the lattice constants of the two materials should not be too different. A thin layer of a material with

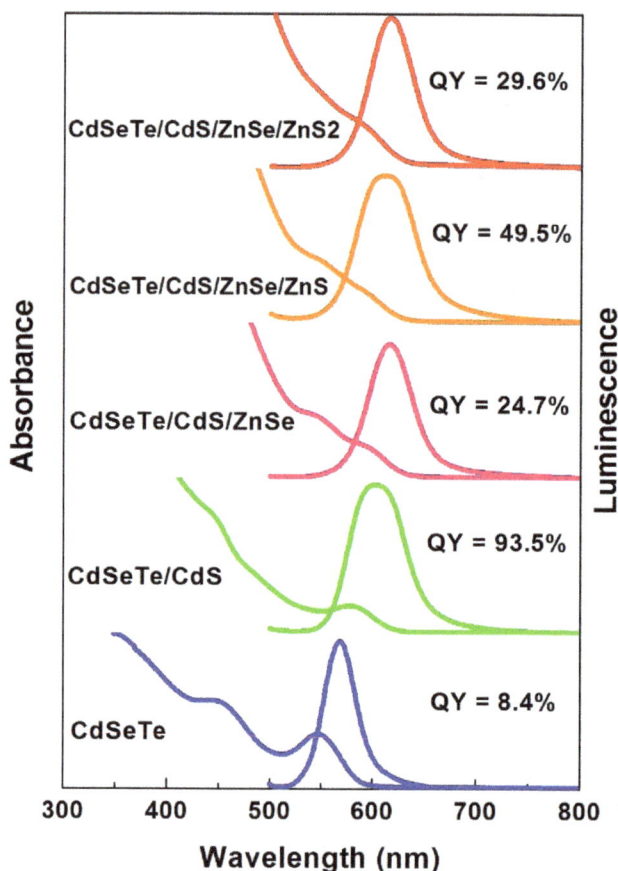

Figure 5.3: UV–Vis absorption and fluorescence emission spectra of the water-soluble alloyed CdSeTe, CdSeTe/CdS, CdSeTe/CdS/ZnSe, CdSeTe/CdS/ZnSe/ZnS (50 mL of ZnS) and CdSeTe/CdS/ZnSe/ZnS$_2$ (100 mL of ZnS) QDs measured in Millipore water. $\lambda_{exc} = 470$ nm, QY: photoluminescence quantum yield (reprinted with permission from Ref. [14]. © (2016) John Wiley and Sons).

intermediate lattice constant can be grown between the two desired shells to reduce strain between the shells.

5.2.3 *Surface capping agents*

Surface modification strategies have enabled the use of QDs in various applications across disciplines. The core or core/shell

QDs are typically prepared with a hydrophobic surface layer derived from the non-polar organic solvents and surfactants utilised during organometallic synthesis, such as 1-octadecene (ODE) and TOPO, which render them insoluble in aqueous media. Surface ligand exchange reactions are therefore employed in order to render the QDs soluble, such as replacement with L-cysteine. The replacement ligands typically have thiol and/or amine functionality (Figure 5.2), as they contain donor atoms (S or N) which possess unshared electron pairs and are thus capable of forming coordinating bonds with metal atoms.

This surface passivation process may have the concomitant effect and advantage of improving the optical properties of the QDs, as so-called "dangling bonds" (arising from the surface atoms having fewer neighbouring atoms than the interior atoms) and surface defects are removed. These dangling bonds and surface defects, which may introduce trap states, have a negative impact on the PL quantum yield of the QDs, because exciton relaxation into localised surface states makes radiative recombination less likely.[15] Stability of the QDs is enhanced by surface capping and aggregation is avoided which would impact on both the absorption and luminescence emission properties of the QDs. It should be noted, however, that it is common for the PL intensity of QDs to be diminished upon conversion of hydrophobic QDs to the water phase.[14]

5.3 Fluorescence Properties of Quantum Dots

Semiconductor QDs have broad absorbance bands, therefore they can be excited using a range of wavelengths. Organic dyes, on the other hand, typically have narrow absorption spectra, which means that they can only be excited within a narrow window of wavelengths. The emission spectra of organic dyes are asymmetric and have a broad red tail. QD emission spectra are symmetric and narrow.

5.3.1 *Effect of quantum dot size*

In solids, the exciton Bohr radius (a_0) describes the extension of excitons (electron–hole pairs), which may range from around 2–50 nm. As the size of a nanocrystal approaches a_0, as in QDs, confinement effects induce changes in the density of the electronic states and in the energy level separation. This leads to an increase of the bandgap (HOMO–LUMO gap) i.e. the electronic excitations shift to higher energy as the particle size decreases (Figure 5.4). In addition, discrete energy levels appear near the band edges. This results in the optical properties of QDs being size dependent and allows for the tuning of the PL properties thereof by choice of composition and size of the QD. QDs are referred to as 0-dimensional objects (0-D), as the exciton is confined in all directions.

Overcoating of core QDs with a shell layer has the concomitant effects of increasing the particle size and thereby leads to redshift in PL emission, as can be seen for CdSeTe; CdSeTe/CdS; CdSeTe/CdS/ZnSe; and CdSeTe/CdS/ZnSe/ZnS QDs in Figure 5.3.

5.3.2 *Effect of semiconductor material*

The band alignment of semiconductor materials utilised in heterogeneous QDs (core/shell QDs) influences the optical properties of the material. Different charge carrier localisation regimes are observed upon photoexcitation, depending on the difference in energy gap between the HOMO and LUMO of the adjoining materials, which are shown in Figure 5.5.

These materials can be combined in core/shell nanoparticles. Thereby, the relative size and the alignment of the band gaps permits the realisation of different properties, as exemplified in Figure 5.6. Type I has the narrow band gap in the core (example: CdSe/CdS). After excitation this results in both charge carriers of the exciton being located in the core. Free valences or surface capping agents at the outside of the shell do not

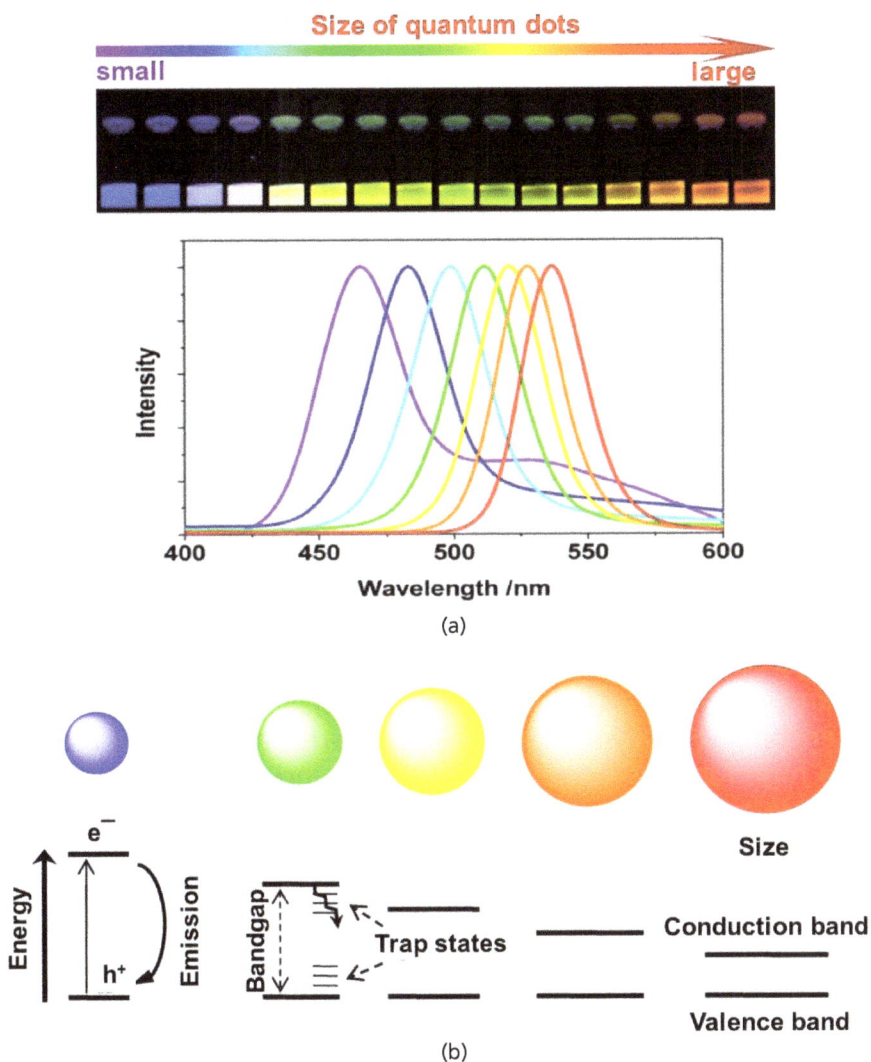

Figure 5.4: (a) Luminescence emission colours from small (blue) to large (red) QDs excited by a near-ultraviolet lamp; with QD sizes typically ranging from ~1 nm to ~10 nm. (b) Creation of exciton upon photon absorption followed by luminescence emission or relaxation through trap states (reprinted from Refs. [16] and [17]).

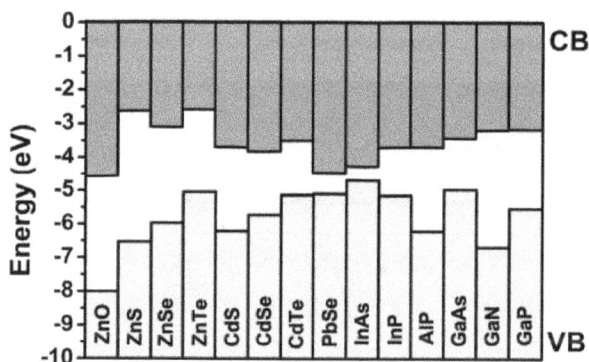

Figure 5.5: The band gap (space between the solid bars) of selected semiconductors, VB: valence band, CB: conduction band (reprinted from Ref. [18]. © (2011), Royal Society of Chemistry).

affect the core properties. The outer shell may contribute significantly to the absorbance, but the charge carriers relax quickly to the core and are therefore not available for transfer to an external reaction partner, i.e. they are protected against quenching. However, they can readily recombine, which leads to a high luminescence quantum yield and makes the particles suitable to serve as fluorescent markers. Changing the size of the core tunes the wavelength of the luminescence.

Swapping core and shell of a type I particle leads to an inverted situation with both charge carriers relaxing to the surface where they become accessible for transfer. This situation is desired for solar water splitting or for solar cells where the charge carriers need to be collected to spend their energy in a chemical reaction or makes it available as electrical voltage. A fast and selective transfer is desired to prevent recombination. For use in titania based dye-sensitised solar cells the energy of the excited electron must be higher than the lower conduction band edge of TiO_2 to allow transfer, but due to the high band gap of TiO_2 the hole cannot transfer and needs an electrolyte shuttle to bring it to the opposite electrode. Certain hydrocarbons are suitable for selective hole capture and transport. In solar water splitting, the H_2O molecule serves to capture holes and release protons. Suitable co-catalysts with selectivity for either hydrogen or oxygen recombination are deposited at the shell surface in order to suppress H and O recombination to water. These applications are discussed in more detail in Sections 7.4 and 7.6.

If the band gap of core and shell are similar, but the one of the core is displaced to higher values, this results in type II particles (example: ZnTe/CdSe). Electrons relax to the shell surface, while the holes remain confined to the core. This provides a charge separated state that prevents recombination of electrons and holes and thus reduces the fluorescence quantum yield. The effective band gap decreases and leads to a strong redshift of the luminescence with respect to absorption and to luminescence of the individual materials. This can be of advantage if luminescence has to be separated from masking due to other spectral features.

As outlined in Section 2.11, doping with guest ions introduces localised defect states which are generally found within

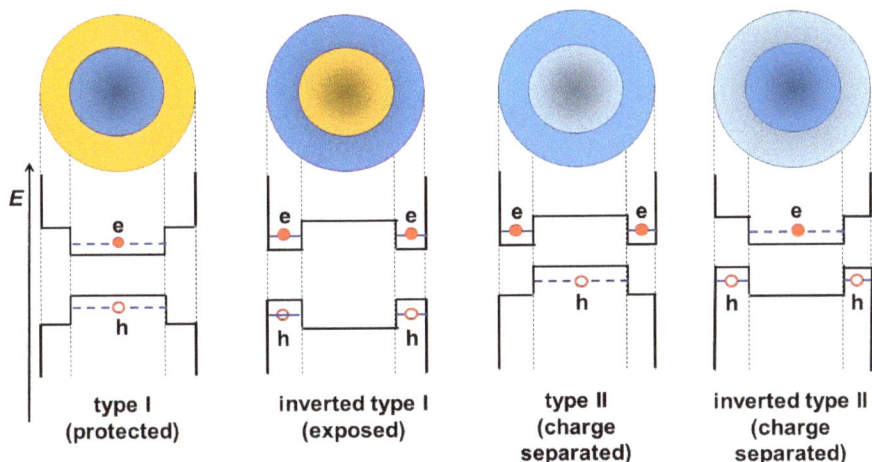

Figure 5.6: Different types of band alignment of core/shell nanoparticles with the main location of excited electron (e) and hole (h) indicated.

the band gap of the semiconductor. This has a significant effect on the electronic properties and therefore also on the optical properties. Special interest has been in dilute doping with manganese in view of applications in magnetic memory for information storage and for spin-based electronics (spin transport electronics, or spintronics) where conductivity is different for spin up and spin down electrons in a magnetic field.

5.3.3 *Photoluminescence quantum yield of quantum dots*

One of the unique optical properties of QDs is their high luminescence quantum yields (PLQYs) and molar extinction coefficients which are 10–100 times higher than that of organic dyes.[19] The PLQY is a relative value based on the ratio of the number of emitted photons to that absorbed, with high PLQYs being desirable. It should be noted that high concentrations of QDs in solution may lead to self-quenching and thereby reduced PLQYs.

Fluorescence quantum yields (Φ_F) can be determined by a comparative method based on luminescence and absorbance intensities:

$$\Phi_F = \Phi_{F(\text{Std})} \frac{F \cdot A_{\text{Std}} \cdot n^2}{F_{\text{Std}} \cdot A \cdot n_{\text{Std}}^2}, \tag{5.1}$$

where F and F_{Std} represent the integrated fluorescence intensity of the QD solution and a reference standard, respectively; A and A_{Std} are the absorbance of the sample and reference standard at the excitation wavelength and n and n_{Std} are the refractive indices of the solvents used for the sample and reference.

For the PLQY determination of QDs in water (for which $n = 1.333$), rhodamine 6G in ethanol ($\Phi_F = 0.95$ and $n = 1.36$) is often employed as a reference standard.[20] Examples of PLQY for CdSeTe; CdSeTe/CdS; CdSeTe/CdS/ZnSe; and

CdSeTe/CdS/ZnSe/ZnS QDs are included in Figure 5.3. In this example, CdSeTe QDs produced a PLQY of 8.4%, which increased dramatically to 93.5% when the CdS shell was passivated on the CdSeTe surface. This increase suggests a well-passivated surface, low defect concentrations and a reduction in the non-radiative recombination rates of the CdSeTe/CdS core/shell QDs.[14]

5.4 Key Points

- A large variety of colloidal semiconductor nanoparticles, called QDs, can be synthesised at a size down to a few nanometres and with a narrow size distribution.
- QDs are generally crystalline and often near-spherical, but other shapes can be engineered. They consist of a core or a core surrounded with one or several shells of different compositions, and generally the surface is terminated by a dense layer of ligands for stabilisation.
- QDs have unique chemical and physical properties including size-dependent fluorescence, high fluorescence quantum yields, narrow spectral bands, independence of emission on the excitation wavelength, and stability against photobleaching. This makes these QDs excellent and highly sensitive fluorescence markers for various applications.
- QDs are electron-delocalised systems and small enough to exhibit quantum confinement effects imparting size-dependent optical properties, thereby allowing for the tuning of their absorption and emission properties by changing the particle size, shape and surface structure of the QDs.
- Semiconductor QDs have broad absorbance bands, therefore they can be excited using a range of wavelengths.
- The band alignment of semiconductor materials utilised in heterogeneous QDs (core/shell QDs) influences the optical properties of the material. Different charge carrier localisation regimes are observed upon photoexcitation,

depending on the difference in energy gap between the HOMO and LUMO of the adjoining materials.

General Reading

- A. D. Yoffe, Semiconductor quantum dots and related systems: Electronic, optical, luminescence and related properties of low dimensional systems, *Adv. Phys.*, 2001, 50, 1–208.
- P. Alivisatos, W. Gu, C. Larabell, Quantum dots as cellular probes, *Annu. Rev. Biomed. Eng.* 2005, 7, 55–76.
- S. V. Kershaw, A. S. Susha, A. L. Rogach, Narrow bandgap colloidal metal chalcogenide quantum dots: Synthetic methods, heterostructures, assemblies, electronic and infrared optical properties. *Chem. Soc. Rev.*, 2013, 42, 3033–3087.
- de Mello Donegà, Synthesis and properties of colloidal heteronanocrystals, *Chem. Soc. Rev.*, 2011, 40, 1512–1546.

References

1. E. Roduner, *Chem. Soc. Rev.*, 2006, 35, 583–592.
2. I. Ekimov, A. A. Onushchenko, *Soviet Phys. Semicond. USSR.* 1982, 16, 775–778.
3. M. A. Reed, R. T. Bate, K. Bradshaw, W. M. Duncan, W. R. Frensley, J. W. Lee, H. D. Shih, *J. Vac. Sci. Technol. B:* 1986, 4, 358–360.
4. O. Adegoke, T. Nyokong, P. B. C. Forbes, *J. Alloys Compounds*, 2015, 645, 443–449.
5. B. Dabbousi, J. Rodriguez-Viejo, F. V. Mikulec, J. Heine, H. Mattoussi, R. Ober, K. Jensen, M. Bawendi, *J. Phys. Chem. B*, 1997, 101, 9463–9475.
6. W. Y. William, E. Chang, R. Drezek, V. L. Colvin, *Biochem. Biophys. Res. Comm.*, 2006, 348, 781–786.
7. H. Zhang, L.-P. Wang, H. Xiong, L. Hu, B. Yang, W. Li, *Adv. Mat.*, 2003, 15, 1712–1715.
8. A. J. Nozik, M. C. Beard, J. M. Luther, M. Law, R. J. Ellingson, J. C. Johnson, *Chem. Rev.*, 2006, 110, 6873–6890.
9. S. J. Rosenthal, J. C. Chang, O. Kovtun, J. R. McBride, I. D. Tomlinson, *Chem. Biol.*, 2011, 18, 10–24.

10. F. D. de Menezes, A. G. Brasil Jr, W. L. Moreira, L. C. Barbosa, C. L. Cesar, R. d. C. Ferreira, P. M. A. de Farias, B. S. Santos, B. S. *Microel. J.,* 2005, 36, 989–991.
11. H. Huang, J.-J. Zhu, *Analyst,* 2013, 138, 5855–5865.
12. S. A. Nsibande, P. B. C. Forbes, *Anal. Chim. Acta,* 2016, 945, 9–22.
13. W. C. Law, K. T. Yong, I. Roy, H. Ding, R. Hu, W. Zhao, P. N. Prasad, *Small,* 2009, 5, 1302–1310.
14. O. Adegoke, T. Nyokong, P. B. C. Forbes, *Luminescence,* 2016, 31, 694–703.
15. C. de Mello Donegá, *Chem. Soc. Rev.,* 2011, 40, 1512–1546.
16. D. Bera, L. Qian, T.-K. Tseng, P. H. Holloway, *Materials,* 2010, 3, 2260–2345.
17. M. F. Frasco, N. Chaniotakis, *Sensors,* 2009, 9, 7266–7286.
18. C. de Mello Donegà, *Chem. Soc. Rev.,* 2011, 40, 1512–1546.
19. K. E. Sapsford, L. Berti, I. L. Medintz, *Angew. Chem. Int. Ed.,* 2006, 45, 4562–4589.
20. J. Lakowicz, *Principles of Fluorescence Spectroscopy,* Kluwer, 2nd ed., New York, 1999.

Chapter 6

Energy Transfer Processes of Excited States

6.1 Types of Energy Transfer

There are many kinds of energy: mechanical (potential and kinetic), heat (also a form of kinetic energy), electrical, magnetic, electromagnetic, chemical. They can be transferred and interconverted. We shall restrict our discussion to energy transfer from states that are a result of excitation in the UV and visible part of the electromagnetic spectrum. Excited states contain one or more electrons with extra potential and kinetic energy. This excess energy can migrate in a crystalline insulator or semiconductor in the form of a Wannier exciton (see Figure 2.7(d)), it can be dissipated fully as heat within the excited molecule or crystallite, or it can be emitted radiatively, as in fluorescence (Figure 2.9).

Alternatively, resonance energy transfer (ET) can take place non-radiatively (without emitting a photon) between a pair of donor (D) and acceptor (A) incorporated inside solids or dispersed in liquid or gaseous phases (Eq. (6.1)).[1] This is to be distinguished from electron transfer (Eq. (6.2)) or hole transfer that occurs in quenching reactions of photo-excited particles.

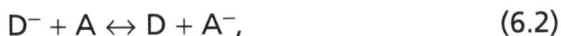

$$D* + A \leftrightarrow D + A*, \qquad (6.1)$$

$$D^- + A \leftrightarrow D + A^-, \qquad (6.2)$$

where the asterisk (*) represents an excited species and the minus sign a negative charge. It is the energy transfer which we want to focus on. If the transfer is between identical states it is reversible and therefore difficult to observe. More often it appears to occur to slightly lower acceptor levels. This provides a driving gradient, and it is accompanied by dissipation of the energy difference ΔE in the form of heat. If ΔE is small, the states are still called approximately in resonance. If it is large compared with thermal energy the transfer becomes directional (non-reversible).

In the case of non-radiative resonance energy transfer we distinguish between Förster resonance energy transfer (named after Theodor Förster) which occurs in non-contact mode over 1–10 nm distance, and Dexter energy transfer, which occurs over distances where the D and A orbitals overlap (<1 nm). Resonance ET plays an important role in upconversion (UC, making one photon from two photons) and downconversion (DC, turning one photon into two photons). DC is also called quantum cutting (QC) and has the potential to rise the external quantum efficiency above 100%.[1] These processes may lead to key technologies in future solar cells.

Often, several transfer processes occur in competition with each other, and in technical applications one attempts to favour the process of interest or to suppress the others.

6.2 Dexter Energy Transfer, Contact Quenching or Sensitisation

Dexter energy transfer (named after David L. Dexter, proposed in 1953[2] is also called collisional transfer, quenching, or sensitisation. Its theoretical treatment involves the transfer of the excited electron from D to A and the (possibly simultaneous) back-transfer of a ground state electron from A to D, so that the overall spin state is conserved (Figure 6.1).

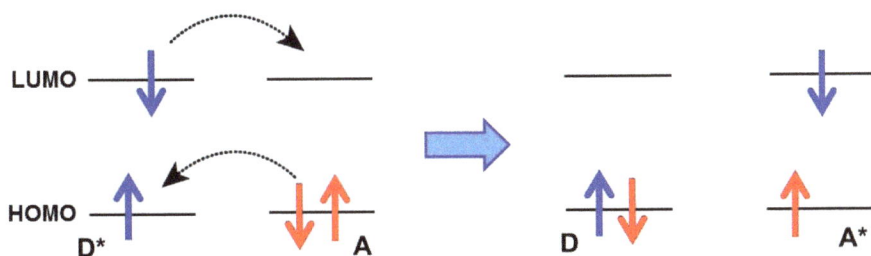

Figure 6.1: Dexter energy transfer from singlet excited state D* to singlet excited state A* by sequential or simultaneous transfer and back-transfer of an electron. The mechanism works analogously for triplet states ^{3}D* and ^{3}A*.

The rate constant k_{ET} for the transfer is assumed to decay exponentially according to

$$k_{ET} \propto J \exp\left(-\frac{2r}{L}\right),$$
(6.3)

where J represents the overlap integral between orbitals on D and A, r is the donor–acceptor distance, and L is the sum of the two van der Waals radii. The exponential decay reflects the similar dependence of the tail of the wave functions. Equation (6.3) holds also for electron transfer, the first step of energy transfer, which is important in the charge separation processes that occur in photovoltaic and photoelectrochemical processes like solar water splitting. In dye-sensitised solar cells the dye is often equipped with carboxyl groups which permits it to be grafted onto the TiO_2 surface in order to enhance the orbital overlap and enhance the rate of electron injection from the excited dye to the titania support.

6.3 Förster Resonance Energy Transfer (FRET)

When a fluorescing chromophore (a donor D, i.e. a dye molecule, a semiconductor quantum dot, or an atom or ion) is excited it deactivates in most cases by fluorescence or by a non-radiative process in which the energy is converted to heat. However, if a second chromophore with slightly lower HOMO–LUMO gap (acting as an acceptor A) and preferentially large extinction coefficient is present within a distance up to about 10 nm, part of the excitation energy of D may be transferred to A. It may be detectable in the form of fluorescence of A (Figure 6.2), but alternatively it decays without emitting fluorescence. The first experimental evidence of such energy transfer which occurs non-radiatively, i.e. without an exchange of photons, and in contrast to the Dexter mechanism also without an exchange of electrons, dates back to 1922 by Cario and Franck.[3] A theoretical explanation and quantitative description was attempted by several scientists. However, the correct dependence of the transfer efficiency that scales with the inverse 6th power of the D–A distance was obtained only after World War II by Theodor Förster.[4] Förster was quite ahead of his time, and a full appreciation of his work had to await recent decades, when FRET became extremely useful as a microscopic ruler in biological systems.

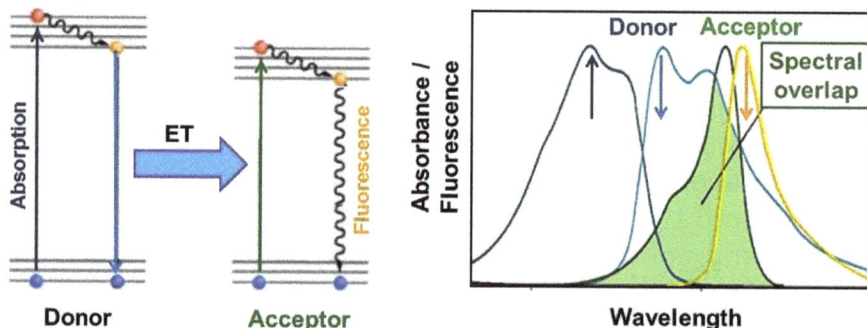

Figure 6.2: Schematic representation of Förster resonance energy transfer from an excited donor (D) to an acceptor (A), which is a strong function of D–A distance. A large spectral overlap between donor emission (blue) and acceptor absorption (green), highlighted in green, is essential for efficient FRET. In the experiment, the wavelengths for excitation and detection of emission of donor and acceptor emission should be chosen for minimal interference and good signal (arrows).

The conditions for efficient energy transfer via the Förster mechanism are[5]:

- A good spectral overlap between donor emission and acceptor absorption bands.
- A large rate constant for fluorescence emission of the donor.
- A large molar extinction coefficient ε of the acceptor.
- A donor–acceptor distance of *ca.* 0.5–10 nm.

The rate constant for FRET energy transfer, k_{FRET}, is given by[6]:

$$k_{FRET}(R) = \frac{1000\,(\ln 10)\,\kappa^2 J_{DA}\Phi_D}{128\pi^5 n^2 N_A \tau_D \,|R_{DA}|^6} = \frac{1}{\tau_D}\left(\frac{R_0}{|R_{DA}|}\right)^6, \qquad (6.4)$$

where J_{DA} is the overlap integral between donor emission and acceptor absorption (Figure 6.2), Φ_D is the fluorescence quantum yield of the donor, N_A is Avogadro's number, R_0 is the Förster radius (see below), τ_D is the fluorescence lifetime of the donor in the absence of an acceptor, and R_{DA} is the distance vector between the electrical transition dipole moments (Eq. (2.8)) of the donor, μ_D, and the acceptor, μ_A. $\kappa^2/|R_{DA}|^6$ describes the theoretically well-known interaction energy of two electric point dipoles of different orientation at a distance vector R_{DA},[6] with

$$k = (\hat{\mu}_D \cdot \hat{\mu}_A) - 3(\hat{\mu}_D \cdot \hat{R}_{DA})(\hat{R}_{DA} \cdot \hat{\mu}_A). \qquad (6.5)$$

n is the index of refraction of the solvent, and its square stands for the relative dielectric constant at optical frequencies. $k = 0$ for orthogonal (perpendicular) dipoles, and maximum for parallel and antiparallel configurations (note that it enters as the square, so the effect is always positive). Equation (6.5) is normally used for the Coulomb interaction of static dipoles, but as we have seen in Section 2.4, the static electric dipole moment of naphthalene is zero for its electronic ground state as well as its excited state, and yet the transition dipole moment may be non-zero. Förster, in his seminal paper,[4] talks about the interaction of moving electron charges, which is consistent with the oscillating charges during excitation. The electric transition dipole moments can be regarded as dipole antennae systems, an expression that is often used in context of light-harvesting systems. Furthermore, if the fluorescing species are embedded in a rigid or highly viscous matrix they retain their orientation over the fluorescence lifetime of the donor. Therefore, when the experiment is done with plane polarised light the polarisation is retained in the observed acceptor fluorescence.[4]

It is often pointed out that Förster energy transfer occurs non-radiatively. Radiative fluorescence of the donor occurs in parallel. But already Förster stressed that these two processes are not the same. In fact, emission and reabsorption is far less efficient than the observed FRET effect.[4] The efficiency E is defined as

$$E = 1 - \frac{I_{DA}}{I_D} = 1 - \frac{\tau_{DA}}{\tau_D} = \frac{R_0^6}{R_0^6 + R^6}. \tag{6.6}$$

I_D and I_{DA} are the donor emission intensities in the absence and in presence of acceptor molecules, respectively, and τ_D and τ_{DA} are the corresponding lifetimes. At the Förster radius R_0 the efficiency is 50%, which means that half of the detected emitted photons come from the donor and half from the acceptor (Figure 6.3). 5 nm, an order of magnitude larger than molecular diameters, is a typical value of R_0. It can be obtained from concentration-dependent (i.e. average distant-dependent) measurements. On this basis, FRET efficiency measurements serve as a spectroscopic ruler for the distances of biological units, e.g. proteins, to which the fluorescing dyes have been attached.

The rate constant for FRET between two molecular fluorophores (fluorescing dyes) decays with R^{-6}, which limits the

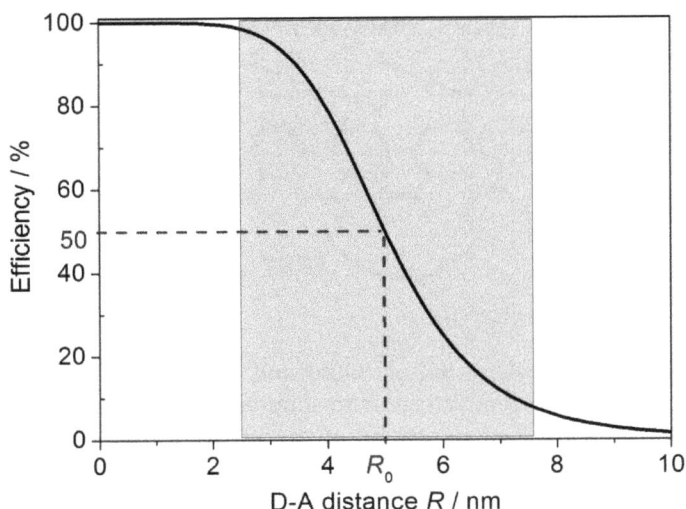

Figure 6.3: FRET efficiency E in % as a function of donor–acceptor distance. The Förster radius R_0 is the distance where fluorescence of donor and acceptor are equal and E amounts to 50% in the absence of radiationless deactivation. The range where FRET is sensitive for a ruler of distances corresponds in favourable cases to $0.5\,R_0 \leq R \leq 1.5\,R_0$ (shaded area).

range of the spectroscopic ruler to *ca.* 10 nm. However, it is known that dye molecules interact with nearby bulk metal surfaces, and this quenches their fluorescence. If the acceptor fluorophore is replaced by a gold nanoparticle that permits the excitation of plasmon resonances it can also act as an acceptor. Due to its near-continuum level structure the decay of the energy follows a distance dependence that scales only as R^{-4} which extends the range of its application as a ruler to *ca.* 70 nm. If both dyes are nanoparticles, a R^{-3} dependence and a further improvement of the useful ruler range up to 100 nm is obtained.[7]

   The orientation factor k in Eq. (6.5) contains the scalar product of the transition dipole moments of the two fluorophores. This implies $k = 0$ for orthogonal (perpendicular) dipole moments, which has the consequence that k_{FRET} is also

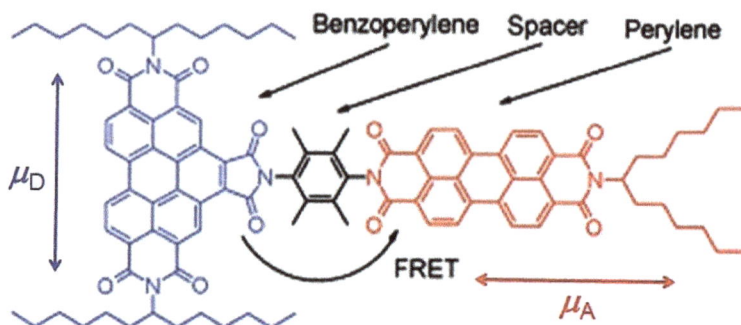

Figure 6.4: Benzoperylene donor (blue) and perylene acceptor (red) connected rigidly via an aromatic spacer to suppress Dexter energy transfer and ensure the perpendicular orientation of the two transition dipole moments (reprinted with permission from Ref. [6]. © (2010) American Chemical Society).

predicted to be zero. This was tested using an arrangement in which donor and acceptor were rigidly connected in a perpendicular geometry (Figure 6.4). The remarkable finding of this careful experiment was an ultrafast FRET that occurred in <10 ps.[6] Dexter energy transfer was excluded by variation of the spacer bridge to which the effect was not sensitive. It is unclear at this point why the orientation dependence of Förster's theory breaks down.

6.4 Excitonic Energy Relaxation

For interchromophore distances of <2 nm, excitonic coupling may become the dominant type of interaction between two or more chromophores. In such a case, energy is not transferred between the excitonically coupled chromophores, but relaxation occurs within the Frenkel exciton manifold (see Section 2.4.8), a much faster process than FRET. When a large number of chromophores interact excitonically, relaxation will occur to the exciton states with the lowest energy. The population of the lowest energy states is determined by the relative orientation of the interacting chromophores' transition dipole moments and the relative strength of their interaction with their environment.

The two limits of dipole interactions are illustrated in Figure 2.8. In the weak-coupling case (Figure 2.8(a)) the interaction between the transition dipole moments is much smaller than the difference in site energies of the two chromophores. As a result, coupling to the phonon bath dominates and the excited-state wave functions are mainly localised on individual chromophores. The spread in absorption bands consequently reflects the spread in site energies. In this regime, excitonic interaction is small, so that no or little exciton splitting occurs and the energy transfer between chromophores occurs via FRET.[8] In the strong-coupling limit (Figure 2.8(b)) the transition dipole interaction dominates the site energy difference and phonon bath interaction. Exciton splitting occurs and the excited-state wave functions are delocalised over the interacting chromophores. Redfield theory[9] provides a realistic description in this regime.[10]

6.5 Stern–Volmer Kinetics and Competition Between Deactivation Channels

The Stern–Volmer relationship[11] is standard for investigations of the kinetics of intermolecular photophysical deactivation. To account for processes like quenching of the excited S_1 state we extend the scheme set out in Section 2.9 by a bimolecular process of a quencher molecule Q with S_1 (for simplicity we subsume intersystem crossing under non-radiative processes since it is kinetically not distinguishable):

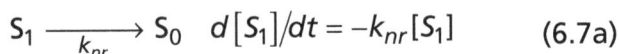

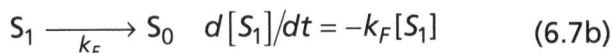

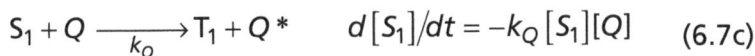

$$S_1 \xrightarrow{k_{nr}} S_0 \quad d[S_1]/dt = -k_{nr}[S_1] \qquad (6.7a)$$

$$S_1 \xrightarrow{k_F} S_0 \quad d[S_1]/dt = -k_F[S_1] \qquad (6.7b)$$

$$S_1 + Q \xrightarrow{k_Q} T_1 + Q* \quad d[S_1]/dt = -k_Q[S_1][Q] \qquad (6.7c)$$

Common quenchers include molecular oxygen, iodide anions or cesium cations (both are heavy atoms which can facilitate intersystem crossing *via* spin–orbit coupling, leaving the fluorophore in T_1), and acrylamide. The choice depends on the accessibility of the fluorophore. Ions are only useful in aqueous environments, acrylamide is relatively large and may not be able to access small pockets of proteins. The small and uncharged O_2 is a very versatile quencher which is assumed to convert to its singlet state and leave the fluorophore in its T_1 state (Eq. (6.7c)). In contrast to chemical reactions, Q is not normally used up; its concentration [Q] therefore remains constant. As in the case of Dexter energy transfer, dynamic quenching requires a collisional encounter between S_1 and Q. The overall kinetics remains first-order, but the apparent rate constant depends on [Q] (pseudo-first order reaction).

In the absence of a quencher, the fluorescence quantum yield is the same as given by Eq. (2.16b), with k_{ISC} assumed zero:

$$Q_F^0 = \frac{k_F}{k_{nr} + k_F} \tag{6.8a}$$

and in the presence of the quencher with rate constant k_Q it becomes

$$Q_F = \frac{k_F}{k_{nr} + k_F + k_Q[Q]}. \tag{6.8b}$$

The ratio of the two expressions is the Stern–Volmer equation:

$$\frac{Q_F^0}{Q_F} = \frac{\tau_0}{\tau} = \frac{k_{nr} + k_F + k_Q[Q]}{k_{nr} + k_F} = 1 + \frac{k_Q[Q]}{k_{nr} + k_F} = 1 + K_{SV}[Q] = 1 + k_Q\tau^0[Q],$$

$$\tag{6.8c}$$

with the fluorescence lifetime in the absence of a quencher, τ^0. Because the fluorescence intensities are proportional to the quantum yields, a plot of the intensity ratios as a function of quencher concentration is expected to give a straight line, with the Stern–Volmer constant $K_{SV} = k_Q \tau^0$ as the slope and with intercept 1.0 (Figure 6.5). Alternatively to the intensity ratio one can plot the lifetime ratio, which is often more robust but is only available from time-resolved experiments. An upper limit of k_Q can be calculated using the Smoluchowski equation for diffusion-controlled reactions. Deviations to lower values can occur if the quenching efficiency per collision with physical contact is below unity (weak quenchers). This changes the slope but not the linearity of the Stern–Volmer plot.

In many cases, however, the plot shows deviation from linearity with either upward or downward curvature.[12] Upward curvature is observed for distance-dependent quenching (e.g. exponentially decreasing with distance, or following the

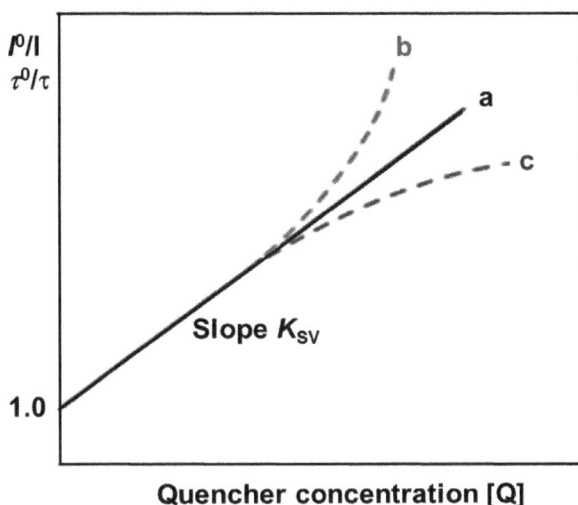

Figure 6.5: Schematic Stern–Volmer plot of the intensity ratio of fluorescence I^0/I and fluorescence lifetime ratio τ^0/τ as a function of quencher concentration in the absence and presence of quencher. (a) represents normal behaviour, (b) and (c) show deviations as discussed in the text.

distance dependence of FRET).[10] Alternatively, it may indicate the presence of so-called dark complexes, for example relatively stable dimers of the fluorophore and quencher which do not exhibit fluorescence. This is called static quenching. It reduces the intensity and causes upward curvature of the intensity ratio plot because the probability of complex formation increases with [Q]. However, it does not affect the lifetime of the remaining fluorescence, and the Stern–Volmer plot of the lifetime ratio remains unchanged. In the presence of nonlinear plots, it is advisable to use both the intensities and the lifetimes for the interpretation.

Downward curvature may indicate the presence of multiple sites of the fluorophore, e.g. when it is bound to a protein. The sites which are more difficult to access are quenched at higher but less likely at lower quencher concentration.

6.6 Key Points

- Energy transfer from excited states can occur radiatively under emission of luminescence, or alternatively non-radiatively in contact between a donor and an acceptor (Dexter type transfer, requiring overlap of donor and acceptor orbitals) or at a donor–acceptor distance of typically 0.5–10 nm (Förster type transfer). Dexter type transfer is also called contact quenching or sensitisation.

- Förster type resonance energy transfer (FRET) requires overlap between the donor emission and the acceptor absorption spectrum.

- FRET serves as a ruler of distances, in particular in combination with single molecule spectroscopy. The rate constant for FRET decays with the sixth power of the donor–acceptor distance, except when nanoparticles are involved.

- Stern–Volmer kinetics involving fluorescence lifetime measurements as a function of the concentration of a quencher can unravel competing deactivation processes.

General Reading

- S. Ramachandra, Z. D. Popović, K. C. Schuermann, F. Cucinotta, G. Calzaferri, L. De Cola, Förster resonance energy transfer in quantum dot-dye-loaded zeolite L nanoassemblies, *Small*, 2011, 7, 1488–1494.
- T. Förster, Zwischenmolekulare Energiewandlung und Fluoreszenz, *Ann. Physik*, 1948, 437, 55–75.
- P. C. Ray, Z. Fan, R. A. Crouch, S. S. Sinha, A. Pramanik, Nanoscopic optical rulers beyond the FRET distance limit: Fundamentals and applications, *Chem. Soc. Rev.*, 2014, 43, 6370–6404.
- M. Wubs, W. L. Vos, Förster resonance energy transfer rate in any dielectric nanophotonic medium with weak dispersion, *New J. Phys.*, 2016, 8, 053037.

References

1. X. Liu, J. Qiu, *Chem. Soc. Rev.*, 2015, 44, 8714–8746.
2. D. L. Dexter, *J. Chem. Phys.*, 1953, 21, 836–850.
3. G. Cario, J. Franck, *Z. Physik*, 1922, 11, 161–166.
4. T. Förster, *Ann. Physik*, 1948, 437, 55–75.
5. S. Ramachandra, Z. D. Popović, K. C. Schuermann, F. Cucinotta, G. Calzaferri, L. De Cola, *Small*, 2011, 7, 1488–1494.
6. H. Langhals, A. J. Esterbauer, A. Walter, E. Riedle, I. Pugliesi, *J. Amer. Chem. Soc.*, 2010, 132, 16777–16782.
7. P. C. Ray, Z. Fan, R. A. Crouch, S. S. Sinha, A. Pramanik, *Chem. Soc. Rev.*, 2014, 43, 6370–6404.
8. T. Förster, *Delocalized excitation and excitation transfer.* In: *Modern Quantum Chemistry.* O. Sinanoglu (ed.), Academic Press, New York, 1965, pp. 93–137.
9. A. G. Redfield, *Adv. Mag. Res.*, 1965, 1, 1–32.
10. V. I. Novoderezhkin, R. van Grondelle, *Phys. Chem. Chem. Phys.*, 2010, 12, 7352–7365.
11. O. Stern, M. Volmer, *Phys. Zeitschrift*, 1919, 20, 183–188.
12. L. Mátyus, J. Szöllősi, A. Jenei, *J. Photochem. Photobiol. B: Biology*, 2006, 83, 223–236.
13. J. Lakowicz, *Principles of Fluorescence Spectroscopy*, 2nd ed., Kluwer, New York, 1999.

Chapter 7

Advanced Applications of Optical Spectroscopy

7.1 Single-molecule Spectroscopy

7.1.1 *Motivation*

Newtonian mechanics treats the trajectories of the considered bodies as a function of time. By doing so, it monitors the coordinates and velocity vectors of each body at all times. Due to the many coupled parameters the number of bodies which can be treated this way is very limited. Chemical and biological samples consist of many bodies reaching the order of Avogadro's number. The description of such samples is based on thermodynamics which drops the many individual coordinates and trajectories and describes only average values over time or over the entire ensemble. According to the ergodic theorem the two averages are the same for random processes. An analogous rationalisation holds for quantum mechanical treatments of molecular systems and extended solids. The Hamiltonian contains all exact coordinates of nuclei and electrons, but for a time-independent Hamiltonian only the time-independent part of the Schrödinger equation is solved. This leads to a time-averaged solution for the electron and nuclear density distributions analogous to thermodynamic ensembles, apart from the quantum-mechanical confinement effect that shows up in standing wave properties and energy quantisation.

Technological advances, notably the ultimate sensitivity of single photon counting detectors combined with confocal microscopy and the availability of well-focused pulsed lasers has led to the development of single molecule spectroscopy over the last three decades. The method that was first carried out in its original version by Moerner and Kador[1] has the following principal advantages:

- It allows monitoring of the behaviour of individual fluorescing molecules, enzymes, quantum dots (QDs) or catalytic reactions that develop fluorescing products, and the interactions with their environment as a function of time. This represents a significant advance over the observation of the probability distributions of ensembles.
- Static and dynamic heterogeneity of observables can be separated on time scales of the fluorescence lifetime.
- Time-dependent processes can be elucidated without the difficulty, or often impossibility, to synchronise these processes in many molecules.
- Statistically rare events can be explored, enabling the possibility to detect new phenomena.[2]

7.1.2 *Principles and crucial requirements of the experiment*

To perform spectroscopy on a single fluorescing unit, two requirements need to be met. Firstly, the fluorescence emission from only one of the objects of interest should be present in the detection volume. This condition is fulfilled by simply using a sufficiently low concentration of fluorophores. In addition, to ensure that no other species contributes to the signal, the fluorophores can be arranged into a monolayer and/or a particular optical configuration can be employed which guarantees that only light from the focal volume reaches the detector. A simple and efficient way to achieve the latter is to add a pinhole behind the microscope. This technique is known as confocal microscopy.[2]

The second requirement for single molecule spectroscopy is that the fluorescence signal from a single fluorophore should be larger than the background, i.e. the signal-to-noise ratio has to be greater than unity for a reasonable averaging time. This condition places severe demands on the experimental setup and the photophysical properties of the sample.[2]

The key factors of the experimental setup can be summarised as follows: well directed and localised laser excitation, nanometre-precision positioning, efficient signal collection, efficient optical filtering of the fluorescence signal and low-noise high quantum-yield detection. The photophysical properties of the excited sample that ensure optimal fluorescence include a large absorption cross-section, weak bottlenecks into dark states such as triplet states (see Section 7.1.3), operation below saturation of the molecular absorption, and a high fluorescence quantum yield. Photochemical stability is improved dramatically by removing freely diffusing molecular oxygen from the sample.[2]

The principle of the confocal setup is displayed in Figure 7.1. A microscope objective with a large magnification serves both to focus the excitation light and to collect the fluorescence or reflected excitation light as a condenser. A dichroic beam splitter reflects the incident light into the objective and concurrently transmits the fluorescence light, thereby efficiently separating these two signals. The distinctive optical component is the small circular aperture, called the confocal pinhole, which allows transmission of the light originating from the focal plane only. The name "confocal" stems from the optical configuration, which is such that the focal plane is imaged onto the pinhole, i.e. the location of the pinhole is optically conjugate to the focal plane.

As a light source for excitation it is convenient to use a laser in the visible range of the spectrum. Many experiments can be done with continuous irradiation, but for time-resolved

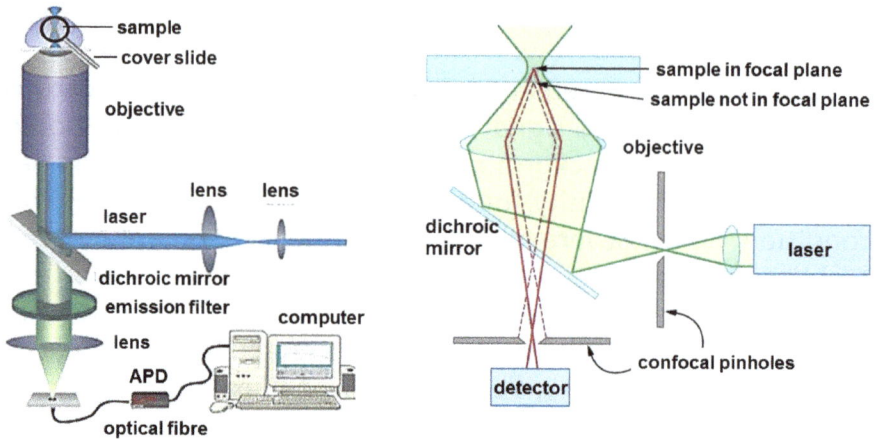

Figure 7.1: Schematic drawing of a fluorescence confocal microscope (courtesy P. Schwille, Ref. [3]) and expanded view of light paths through the optical setup (reprinted from Ref. [4]).

(Section 3.5.3) and autocorrelation experiments one needs a pulsed laser. In an autocorrelation experiment, one measures the time lag between excitation and appearance of the corresponding fluorescence photon (see Figure 7.3). Most experiments are done better with non-polarised light, but rotational diffusion experiments require a polariser or a plane-polarised source.

As a result of the large refractive index mismatch between a typical glass microscope coverslip and the aqueous medium inside the sample cell a significant fraction of the exciting light is reflected back by this interface. This back-reflected light would severely obscure the detection of fluorescence if it were not filtered out by a sharp long-pass filter with its edge little above the exciting wavelength. A 2-dimensional piezo stage allows to position the sample stage and to execute lateral scans. For high-efficiency single-photon detection, either a single-photon silicon avalanche photodiode (APD), a high-level scientific complementary metal-oxide-semiconductor (sCMOS) camera or a low-noise back-illuminated charge coupled device (CCD) camera can be used.[2]

The requirement of having a single molecule in the detection volume of typically 1 femtolitre translates into a concentration of roughly 2 nM of fluorescing probe molecules. In favourable cases this value may be extended down to 1 pM, but enzyme reactions require higher concentrations (μM to mM), whereas diagnostics and biosensing require significantly lower concentrations (<pM). Reduction of the observation volume, confinement of molecules, temporal spacing of fluorescent signals and smart experimental designs help to reduce the higher concentrations of labelled species.[5]

7.1.3 *Intermitting fluorescence activity (blinking)*

A fluorophore is said to "blink" when its fluorescence intensity repeatedly drops to zero and then comes back to normal. In contrast, photobleaching of organic fluorophores is an irreversible switching-off process, primarily by chemical interaction of one of its long-lived excited states with molecular oxygen, or by radical formation in reactions of its ground state with O_2. Blinking appears universal for all fluorophores exposed to interactions, the only exceptions being emitters that are well-shielded from fluctuations in the local environment such as nitrogen vacancy centres in diamond and especially engineered QDs.[6] A typical example of blinking is shown in Figure 7.2. The distribution of time periods where fluorescence is on (τ_{on}) and the ones where it is off (τ_{off}) both obey a universal *power law*, $P \sim 1/\tau^{1+\alpha}$, of the type as they occur for many natural phenomena.[7] Common causes of blinking are discussed below.

Triplet blinking: Fluorescence is emitted quite efficiently from the singlet excited state, but a small fraction of excited states undergo intersystem crossing (Section 2.6) to the triplet state, which does not fluoresce and is therefore called a dark state. In the presence of O_2 which is an efficient triplet quencher the molecule returns quickly to the ground state and becomes available for the next absorption-fluorescence cycle so that the excursion over the triplet state remains unnoticed. However, if

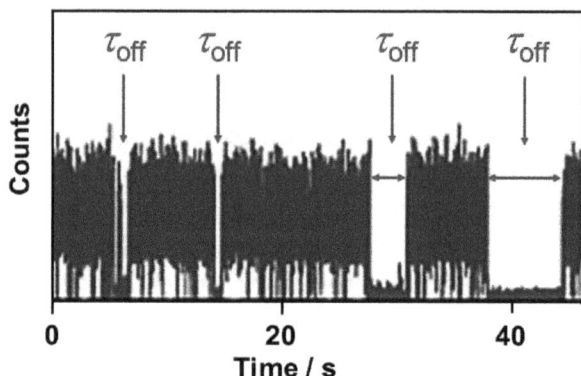

Figure 7.2: Schematic example of a blinking molecule or quantum dot with on and off periods during continuous irradiation with exciting light. No photodegradation is visible on the time scale of observation.

oxygen is removed to suppress photobleaching, this increases the triplet lifetime often by orders of magnitude so that a significant fraction of fluorophores becomes trapped in the dark state and is unobservable for some time corresponding to the natural triplet lifetime. It then returns to the ground state and enters the absorption-fluorescence cycles again. Because blinking is undesired for most applications, b-mercaptoethanol that does not cause photobleaching is used as an equally efficient triplet quencher because of the heavy Hg atom.[6]

Redox blinking: Excited states are more prone to oxidation (giving off the excited electron) as well as reduction (accepting an electron into the hole in the former HOMO) than the ground state. Both leads to radical ions in the ground state, and they do not fluoresce. However, if the fluorophore in its ground state does not undergo redox reactions in the same environment then the system will return to the initial state where it is available for fluorescence again. Depending on the redox potentials of the reaction partners, adding oxidants such as oxygen or methylviologen, or reducing agents such as ascorbic acid or ferrocene derivatives, can induce redox blinking with longer off-periods than for triplet blinking.[6] On the other hand, specific redox systems may be added to neutralise the radical ions and restore stable fluorescence.

Reversible photochemistry-induced blinking: Not only redox reactions lead to blinking, simple (often unimolecular) reversible transformations have the same effect if the transformed species is fluorescence inactive.

Fluorescent proteins: These are proteins that contain one or more fluorophores, such as photosynthetic light-harvesting complexes (Section 7.5) or the green fluorescent protein. When they are used for the purpose of fluorescence labelling (Section 7.2), they should be ideally small (*ca.* 25–30 kDa).[6] A number of sources causing blinking on timescales of nanoseconds to minutes have been identified. Besides the common mechanisms discussed above the origin can also be proton transfer or conformational dynamics, normally involving cis–trans isomerisation of the chromophore, in particular when this leads to non-planarity of its conjugated π-system. Based on this cis–trans isomerisation, some of the chromophores can be actively switched on and off by optical pulses at suitable wavelengths.[6] It was recently shown that protein supercomplexes as large as 5–10 MDa, which bind almost 400 chromophores, undergo fluorescence blinking.[8] Photosynthetic light-harvesting complexes show the intriguing behaviour that blinking is controlled by the protein complex to serve a functional purpose: when the organism needs photoprotection (Section 7.5.4), the dynamic equilibrium between "on" and "off" states shifts to the latter, i.e. blinking occurs more frequently and the complex remains in a dark state for longer times.[9] Blinking in photosynthetic complexes results from small protein conformation changes, which open thermal energy-dissipation pathways by increasing energy transfer to dark carotenoid S_1 states or to charge-transfer states between chromophore pairs.[8,9]

Quantum dots: These are significantly larger than organic chromophores and even than enzymes, in particular once they are made water-soluble and linkable to biological systems by surface grafting. They are therefore not suitable for

Förster resonance energy transfer (FRET) studies but ideal for particle tracking for example at the cell surface or on DNA.[6] Blinking can be successfully suppressed with thiols in solution. Photoluminescence can be quenched by supplying charges which undergo non-radiative recombination with the exciton hole in the core. This recombination can be intercepted when electron-accepting surface sites are present. Both mechanisms can be electrochemically controlled.[10]

Graphene oxide: Unlike Graphene, graphene oxide has a band gap that gives rise to strong fluorescence in the visible spectrum. Unexpected blinking during photoreduction of graphene oxide to reduced graphene oxide is attributed to the redistribution of carbon sp^2 domains.[11]

Fluorescence correlation spectroscopy: If one wants to focus on the behaviour of one particular species, preferably the only one in the detection volume, one performs an autocorrelation analysis. The experiment measures the elapsed time of the corresponding fluorescence photon after the chromophore has been excited with a short laser pulse. Such an excitation-detection event is called a correlated event. Many such successive events on the same molecule are plotted in a time histogram, called the autocorrelation plot, in which a time bin counts the number of events with the same elapsed time after excitation. For such a plot, one normally expects a mono-exponential decay curve that indicates the lifetime of the excited state (i.e. the fluorescence lifetime). In reality, one often observes deviations from this mono-exponential behaviour. For example, in case of blinking, the fluorescence photon appears delayed because the excitation energy was stored during the off-time. In other cases the fluorescence photon does not appear at all within the observation window so that it does not lead to a count.

Figure 7.3 illustrates schematically on what time scale a number of processes lead to deviations from mono-exponential behaviour. Due to the logarithmic time scale the first section, called antibunching, appears enhanced. It represents various deviations from Poisson counting statistics, among it the time

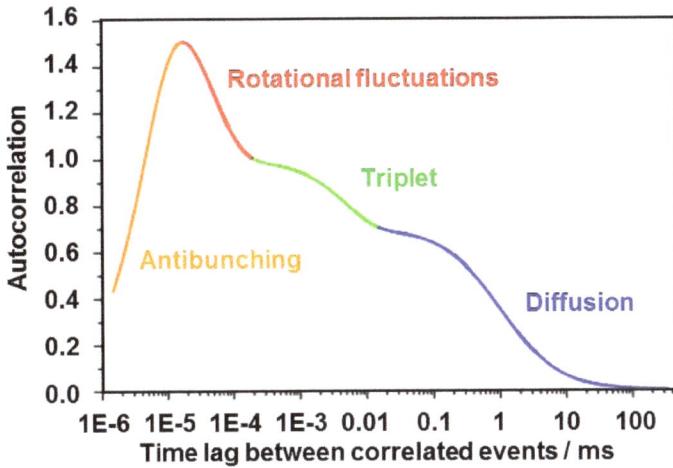

Figure 7.3: Time scale of various processes monitored by autocorrelation analysis. Note the logarithmic time scale. A mono-exponential decay curve would result in a straight line with negative slope in this plot (courtesy P. Schwille, Ref. [3]).

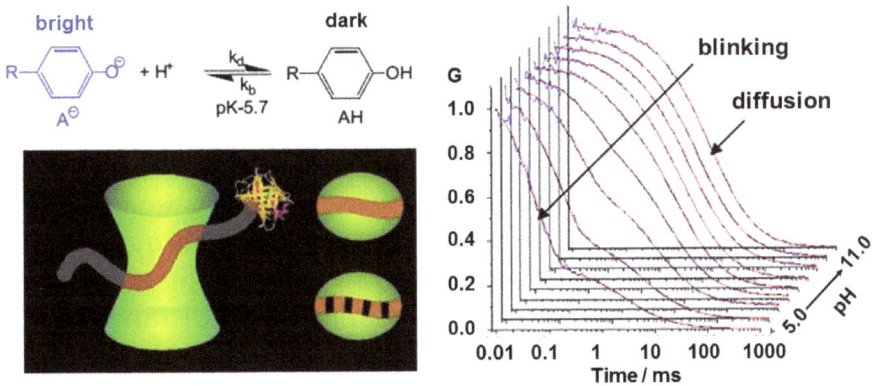

Figure 7.4: Reaction scheme showing the transition from the deprotonated bright to the protonated dark state of the given dye. The autocorrelation curves show qualitatively different behaviour at different pH (courtesy P. Schwille, Ref. [3]).

lag before the first fluorescence photon can appear because the excited chromophore first has to undergo vibrational relaxation to reach the vibrational ground state of the electronic excited S_1 state (Figure 2.9). It is not of interest here. Rotational fluctuations or rotational diffusion reduce the probability that the photon from excitation is emitted back in the direction of the incoming plane-polarised light. Off-time during triplet blinking extends the fluorescence lifetime as discussed above. Translational diffusion takes the chromophore out of the focus of the observation volume.

Figure 7.4 shows an experimental example of changing autocorrelation curves G under the influence of protonation of the dye. The bright state at pH 11 is only affected by diffusion, while the state at a pH slightly below the pK_a value of the dye shows diminished fluorescence events because the protonation leads to a spectral shift of the absorption band from 488 to 400 nm so that the dye can no longer be excited with the light source that was used here. Protonation has the same effect as triplet blinking.[3]

7.1.4 *Instructive examples of applications*

Fluorescence lifetime spectroscopy: Double or even multiple exponential decays in lifetime spectroscopy can indicate that there is more than one chromophore in the observation volume.

Single-molecule Fluorescence Resonance Energy Transfer (smFRET): The energy of an excited singlet state can be transferred via a dipolar coupling mechanism over distances from a donor to an acceptor dye (Section 6.3). Since FRET is a very sensitive function of distance in the range of 0.5–10 nm it can be used at the single molecule level to monitor dynamic intramolecular conformational changes of biomolecules like enzymes or other proteins. Quantum dot nanoparticles are too large and not suitable on this scale, but small organic fluorophores which are covalently bound to the biomolecule and have large extinction coefficients and an overlap between the donor fluorescence and the acceptor absorption spectrum are ideal.[6]

Single particle tracking: Particle tracking using covalently bound organic dyes or semiconductor QDs can access distances from sub-nanometre scale to hundreds of nanometres and may be used for measurements of translational diffusion by analysis of mean-square-displacement analysis[12] or end-to-end distances of DNA strands.[13] Such measurements can occur in various environments, even inside biological cells.

Fluorescence anisotropy measurements: Absorption and fluorescence of organic chromophores are governed by the transition moment, an oscillating vector property with a direction that is related to the orientation of the molecule (Section 2.4.5). If the excitation is carried out with plane-polarised light, absorption will be a maximum when the molecular transition moment is parallel to the plane of polarisation, and a minimum (nominally zero) perpendicular to this direction. If the chromophore resides in a rigid environment so that it keeps its orientation over the fluorescence lifetime, the fluorescence will be polarised parallel to the polarisation of the exciting light, independent of the time span between excitation and emission of the corresponding photon. However, if the chromophore changes its orientation (i.e. if it undergoes rotational diffusion) this causes the polarisation of the fluorescence to deviate, and for long time spans it will be independent of the polarisation of the incident light, i.e. the fluorescence approaches an unpolarised state. If all chromophores experience the same environment at all times, the measured correlation spectrum of an ensemble is expected to be mono-exponential, but if this is not the case the decay curve will be distorted and represented by a stretched exponential.

What are such studies good for? Quite unexpectedly, it was found that fluorescing probe molecules embedded in supercooled glass-forming liquids lead to stretched exponentials. Thus, although these samples look solid they represent a frozen liquid-like state that is macroscopically isotropic.

The non-exponential behaviour suggests that the environment that is probed by the different molecules or at different times is heterogeneous. It provides a picture of the length and time scales associated with dynamic heterogeneity.[14]

Spherical QDs are isotropic and therefore not suited for rotational diffusion studies, but non-spherical, rod-like crystallite possess also an oriented transition dipole moment.

Site-specificity of fluorogenic catalytic reactions: It is of great interest for a mechanistic understanding to learn where the exact site of catalytic activity on an enzyme or a heterogeneous catalyst crystallite is located and how it evolves with time. If a catalytic reaction produces (or uses up) a fluorescing reactant this leads to enrichment (or depletion), which can be monitored *in situ* with high spatial and temporal resolution by single molecule spectroscopy.[15] This contrasts with many other highly sophisticated analytical methods which often remain static, provide only ensemble averages, or require even high vacuum which departs significantly from the conditions under which the reactions typically occur. Single molecule fluorescence spectroscopy can be carried out at room temperature and in liquid or cellular systems, giving evidence of perhaps unexpected dynamic disorder and spatial heterogeneity on a single turnover basis that was previously masked in ensemble averages. Plotting the number of correlated photons arriving as a function of time after the corresponding excitation pulse, one might normally expect a mono-exponential decay curve that indicates the lifetime of the excited state (i.e. the fluorescence lifetime). This is indeed the case under homogeneous conditions. However, in case of dynamic disorder, when a waiting time due to environment-dependent off-times that can span several orders of magnitude, as described in Section 7.1.3, is involved this leads to stretched exponentials. Such behaviour is found with enzymes that show conformation-dependent fluorescence activity.[15]

Figure 7.5 shows the photocatalytic reduction of the non-fluorescent boron-dipyrromethene compound 3,4-dinitrophenyl-BODIPY (DN-BODIPY) to the highly fluorescent

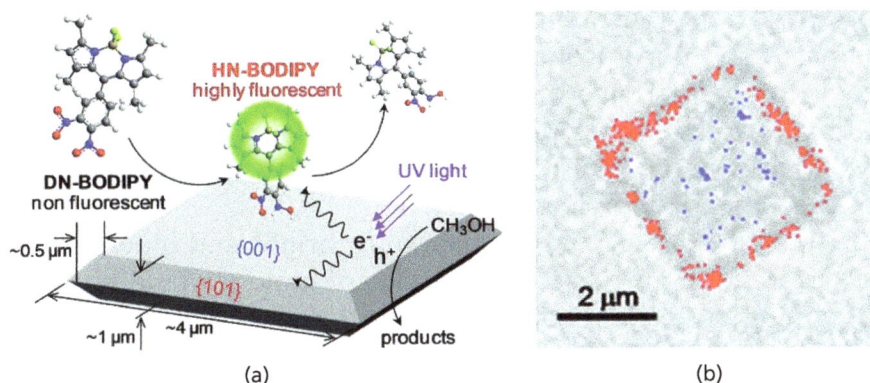

Figure 7.5: (a) Photocatalytic generation of fluorescent HN-BODIPY from non-fluorescent DN-BODIPY over a TiO_2 crystal. (b) Image of immobilised TiO_2 crystal under 488 nm laser and UV irradiation. The blue and red dots show the location of fluorescence bursts on the {001} and the {101} facets of the crystal (reprinted with permission from Ref. [16]. © (2011) American Chemical Society).

4-hydroxyamino-3-nitrophenyl-BODIPY (HN-BODIPY) on the {001} and {101} crystal facets of titania. It is immediately clear that the reaction is favoured on the {101} facet. A kinetic analysis revealed the reaction rate constants to be 0.59 μM {001} and 1.4 μM {101}, respectively. The compelling difference was attributed to a face-dependent electron trapping probability.[16]

Many further applications have been discussed in a recent review by Mörner.[17]

7.2 Labelling in Biology

Optical microscopy is one of the key biophysical methods in current life sciences research. Fluorescence microscopy, in particular, allows biological processes to be studied as they occur in space and time, both on the cellular and molecular level.[18] It is by far the method most often applied and is the dominant analytical approach in a large variety of schemes in the fields of medical testing, biotechnology and drug discovery.[19] Fluorescent proteins from the green fluorescing protein (GFP) and other families have become indispensable and highly sensitive marker tools for imaging live cells, tissues and entire organisms.[18] The Nobel Prize in Chemistry 2008 was awarded to O. Shimomura, M. Chalfie, and R. Y. Tsien "for the discovery and development of the Green Fluorescing Protein".

Apart from GFP that is remarkably bright and glows under UV irradiation, intrinsic fluorescence of biological molecules is of limited use for imaging applications of high resolution because it is weak and non-specific. Therefore, taking advantage of the selective ligand binding sites on proteins, a large number of fluorescent labels has been developed and are commercially available,[20] among them synthetic organic dyes,[21,22] inorganic nanocrystal QDs, special fluorescent proteins (FPs), antibodies,[23] and in particular also the NV defect-containing biocompatible nanodiamonds (Sections 2.10 and 7.7). FPs have the key advantage that they can be genetically encoded so that they are produced by the organisms inside the cells so that no additional label has to be introduced. An extensive list of available FPs and their properties has been compiled by Nienhaus[18] and elsewhere.[24] Advantages and disadvantages of QDs as fluorescing labels are compared with organic dyes by Resch-Genger et al.[25]

Living cells and other biological samples are heterogeneous environments which cause interference by a variety of fluorescing and scattering functional units. Redshifted fluorophores are generally preferred in experiments with living tissues and organisms because of lower phototoxicity and increased penetration depth of the exciting light. Moreover, red emission can readily be separated from the self-fluorescence of biological samples in the green spectral region.[18] Nevertheless, a high brightness and photostability are of utmost importance in the presence of the unavoidable background. Advanced deconvolution techniques as outlined in Section 3.7 are essential for reliable, quantitative interpretation of the results.

Ensemble FRET has long been used to measure distances in the 0.5–10 nm range (Section 6.3). This method provides an ensemble average of the distance. In contrast, dynamic distance changes on a millisecond time scale are obtained for a biopolymer like DNA by focusing on single donor–acceptor pairs at a time when the fluorescent donor and the corresponding acceptor are attached to two different sites on the polymer (Figure 7.6). This single pair FRET (spFRET) allows the study of (i) fluctuations and stability of macromolecules, (ii) macromolecular folding and unfolding using a hairpin conformational model, and (iii) enzyme structural changes during catalysis.[26] Conducting the experiment with polarised excitation and detection allows studying changes in the relative orientation between donor and acceptor (time resolved Fluorescence Polarisation Anisotropy measurements, FPA). Intermolecular interactions can be investigated when acceptor and donor are attached to different molecules. The distance dependence of two GFP type probes on calmodulin, a Ca^{2+} binding protein, was shown to be a sensitive reporter of Ca^{2+} concentration in cells.[27]

Figure 7.6(b) demonstrates that the FRET efficiency is high for the folded and low for the denatured case. The bimodal distribution in between indicates that the denaturation coordinate has two free energy minima. Figure 7.6(c) shows that the wild type of the enzyme needs a higher concentration of the denaturation agent than the mutant. The wild type must therefore be more stable.

Cells employ a variety of linear motors, such as myosin, kinesin, and RNA polymerase, which move along and exert force on filamentous structure.[28] There are quite a few molecular machines that rotate, but undoubtedly the most striking of all is the adenosine triphosphate (ATP) synthase.[26] ATP is the universal energy currency of the cell. When it is hydrolysed to adenosine diphosphate (ADP) and inorganic phosphate (P_i) its biochemical standard free energy $DG°(ATP)$ of -30 kJ mol^{-1} can

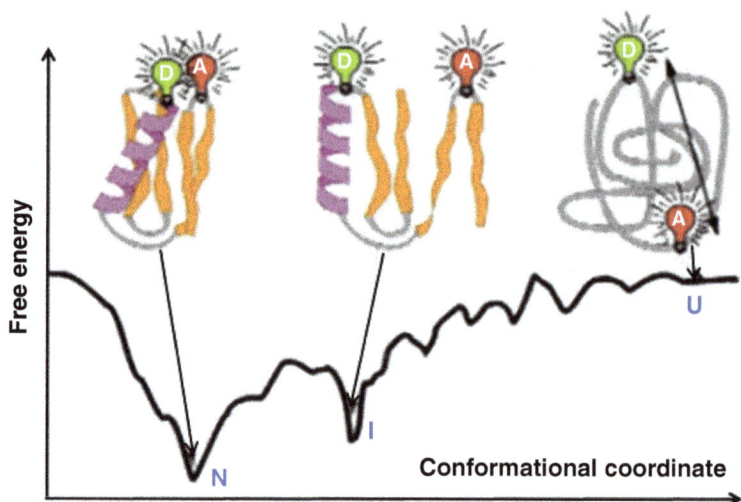

(a)

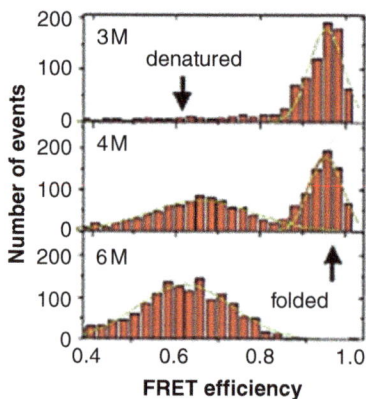

(b)

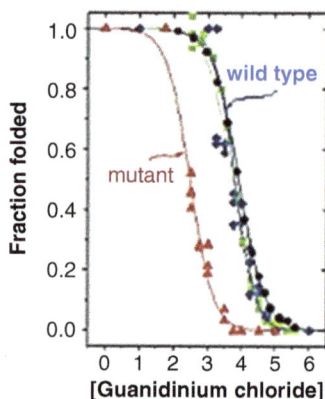

(c)

Figure 7.6: (a) Schematic view of free energy landscape of the conformation of a macromolecule with the folded state N, and intermediate state I and an unfolded state U. The conformational dynamics is monitored by attaching a fluorescent donor to one end and an acceptor to a different position on the polymer and measuring the dynamics of the distance between the two fluorophores using single molecule FRET and of their orientation by means of fluorescence depolarisation. (b) Single molecule FRET histogram showing the FRET efficiency for three different concentrations of a denaturant. (c) Folded fraction from ensemble and single molecule FRET of two variants of an enzyme as a function of denaturant concentration with sigmoid fit from which the distance is derived. (reprinted with permission from Ref. [26]. Copyright (2000) Springer Nature).

be used for a variety of chemical reactions.[29] Under physiological (non-standard) conditions the free energy gain is even higher, up to -60 kJ mol^{-1}. ATPase acts as a rotary stepper motor with a central rotor of ≈ 1 nm diameter that turns in a stator barrel of ≈ 5 nm diameter (Figure 7.7). Unlike the conventional electrical motors which run on electron currents the ATPase runs on a proton current, driven by the pH gradient over the membrane. It was demonstrated by attaching a fluorescent actin filament to the rotor as a marker, and using time resolved single molecule Fluorescence Polarisation Anisotropy measurements (smFPA), that in the presence of ATP instead of a proton

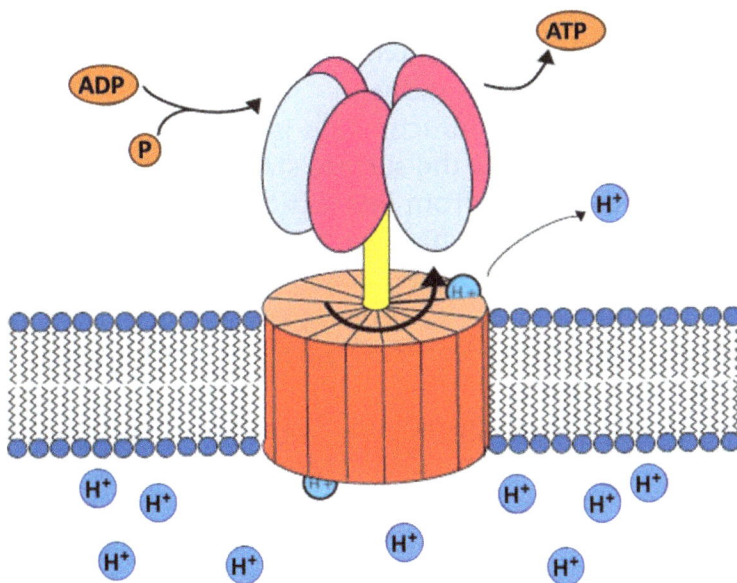

Figure 7.7: Schematic image of the ATPase with the membrane-embedded rotating F$_0$ unit (red) that couples the transmembrane proton flux to ATP synthesis/hydrolysis. It is powered by the pH gradient over the membrane.

Note: Based upon material developed by the Materials Research Science and Engineering Center on Structured Interfaces at the University of Wisconsin–Madison with funding from the National Science Foundation under award number DMR-1720415. Any opinions, findings, and conclusions or recommendations expressed in this report are those of the authors and do not necessarily reflect the views of the Foundation.

gradient the filament reversed and rotated with more than 100 revolutions per second in 120° steps in the anticlockwise direction when viewed from the membrane side.[28]

7.3 Environmental Luminescence Sensors Based on Semiconductor Quantum Dots

Semiconductor QDs have broad absorbance bands; therefore, they can be excited using a range of wavelengths. QD emission spectra are symmetric and narrow, with the maximum occurring at a longer wavelength than that of the absorption spectrum. The spectrally narrow emission signal provides selectivity when employing these materials as fluorescence sensors for environmental applications, which include both inorganic and organic analyte determinations.

Sensing with semiconductor QDs is based on the sensitivity of their luminescence to the surface state of the nanoparticles. Thus sensing can result from interaction (chemical or physical) with the target analyte, which can lead to either photoluminescence enhancement or quenching. These nanosensor-based analytical techniques are of great interest due to their potential for portable, rapid and real-time determinations of environmental pollutants.[30]

The basis of most QD sensing assemblies, however, is not direct interaction with the QD surface, but rather energy flow between the QDs and analyte molecules. FRET occurs when energy absorbed by a donor (usually the QDs) is transferred to a nearby acceptor species. The efficiency of this non-radiative energy transfer depends on (i) the distance between the FRET pair (donor and acceptor), (ii) spatial arrangement or orientations of the pair, (iii) the spectral overlap of the donor emission and acceptor absorption, and finally (iv) the fluorescence lifetime of the donor should be long enough for FRET to occur. Evidence of FRET is seen when there is a decrease in the fluorescence and excited state lifetime of the donor while the fluorescence of the acceptor increases. Therefore QDs are suitable as sensors through the FRET mechanism because of their broad absorption spectra, high quantum yields and long fluorescence lifetimes. It should be noted that while a silica shell or encapsulation of QDs by a polymer/phospholipid may provide good water solubility and photoluminescence quantum yield, they may significantly increase the overall QD particle size and thereby restrain the application of QDs in FRET processes because of the resulting limitation in terms of proximity of acceptor and QD core.

Besides the use of QDs themselves as fluorescence sensors, they may also be combined with other materials as nanocomposites. An example is the embedding of QDs in nanosheets of graphene oxide (GO), to enhance the interaction of the target analyte (in this case phenanthrene, a polycyclic aromatic hydrocarbon) with the QDs (Figure 7.8). For this sensor, a PL enhancement effect was found for phenanthrene, as the intensity of the emission of this QD–GO probe at 633 nm was progressively enhanced as the concentration of phenanthrene was increased.

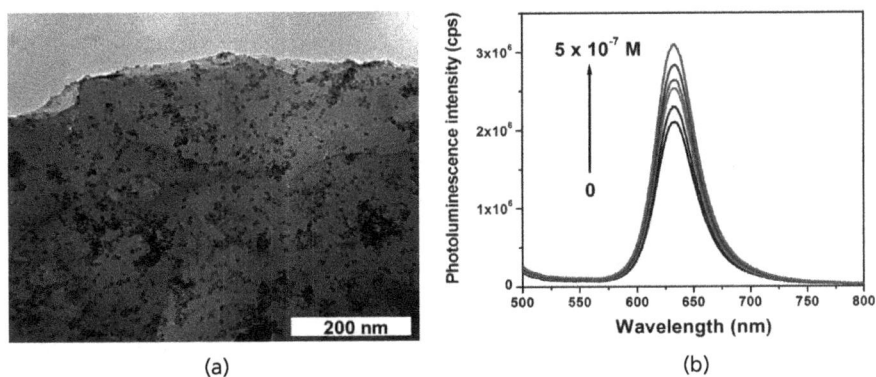

(a)   (b)

Figure 7.8: L-cysteine-capped CdSeTe/ZnSe/ZnS QD-graphene oxide nanocomposite used for fluorescence sensing of phenanthrene. (a) TEM image. (b) Fluorescence detection of phenanthrene at increasing concentration corresponding to a steady enhancement in PL signal on excitation at 470 nm. [Phe] = 0, 1×10^{-7}, 2×10^{-7}, 3×10^{-7}, 4×10^{-7} and 5×10^{-7} mol L^{-1} (λ_{exc} = 470 nm) (reprinted with permission from Ref. [32]. © (2016) Elsevier).

Phenanthrene alone does not absorb above 300 nm, in particular not at the excitation wavelength of 470 nm, nor does it luminesce at 633 nm.[31] The QD luminescence enhancement was attributed to adsorption of phenanthrene on GO, accompanied by $\pi-\pi$ interaction between the two systems. The simplicity, low cost and low luminescence optical density of this sensor (0.19 μg L^{-1} = 1 nM) demonstrated its potential for application as a highly sensitive screening tool in environmental water pollution monitoring.[32]

"Host" molecules which contain a binding site which is highly specific for a "guest" analyte may also be employed in fluorescence sensing applications (so-called "host-molecule-coated" QDs). Cyclodextrins, crown ethers and porphyrins are examples of host molecules that have been reported, where the optical sensors combine the advantageous highly luminescent properties of QDs with the selectivity which the host molecules impart.[33] Use of antibodies, enzymes and nucleic acids as selective moieties in nanomaterials for environmental sensing applications has also been reported.[30]

QDs have also been employed as luminescent probes in the detection of inorganic analytes, typically metal cations. As with organic analyte determinations, fluorescence quenching and enhancement effects account for the majority of the sensing mechanisms. The decrease in fluorescence intensity may be due to adsorption and formation of species with lower solubility than the original QDs, thereby causing their precipitation or agglomeration. The interaction of the target analyte with the QD surface may, however, induce enhancement of fluorescence due to surface passivation, thus minimising surface defects and non-radiative electron/hole recombination pathways.[34] The analyte may form a complex with the ligand on the surface of the QDs, resulting in fluorescence enhancement. Capping agents may also be involved in energy- and charge-transfer processes, altering the fluorescence properties of the material.

7.4 Light Harvesting in Solar Cells

7.4.1 *Introduction*

Solar cells harvest solar radiation that reaches our planet and convert part of it to photovoltaic electricity. The solar flux density is described by what has become known as the solar constant, 1367 W m^{-2}, measured outside the earth atmosphere. In fact, it is not a constant; rather, it is the average of the energy flux per time that illuminates our planet, and it oscillates due to the elliptical orbit of the earth. Moreover, it depends on the solar activity that follows several multi-year cycles.

The solar energy flux that arrives at sea level is about 75% of the flux above the atmosphere, i.e. roughly 1 kW m^{-2}. The attenuation factor that is commonly used is termed air mass 1.5 (AM1.5). It accounts for the fact that the path length travelled by light through the atmosphere at various angles is roughly a factor of 1.5 larger than the direct vertical distance. The flux depends on the angle about which the earth rotates, and it is furthermore attenuated by absorption mostly by water, oxygen and carbon dioxide (Figure 7.9). Light clouds already reduce the solar flux by 50%.

The sun sends us no bill for the energy that it provides. And solar energy is abundant. It sends us the annual world energy consumption in one hour. This is about a factor of 9,000 more than what the world population expends.

7.4.2 *Direct absorption across the band gap*

When photovoltaic cells absorb a photon, an exciton is formed. The crucial process is the separation of the electron and the hole of this exciton and transporting them to two different electrodes.

Most metals are of shiny, greyish appearance, meaning that apart from some reflection there is relatively unselective

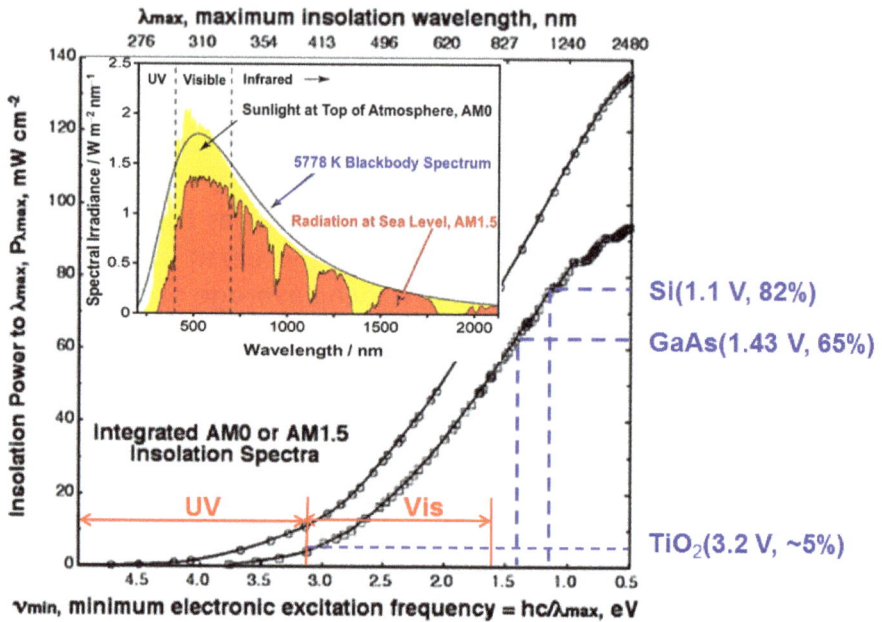

Figure 7.9: Inset: solar spectral irradiance [W m^{-2} nm^{-1}] above the atmosphere (yellow) and at sea level (red, assumed to correspond to AM1.5 — see text). The indentations are absorption bands of atmospheric H_2O, CO_2 and O_2. The solid line corresponds to the calculated black-body radiation for a temperature of 5778 K. The main image gives the integration of the yellow and the red curves up to a running threshold value. 82% of the solar energy spectrum lies above the band gap of Si (1.1 eV), 65% above that of GaAs (1.43 V), but only around 5% above that of TiO_2.

absorption over the entire visible spectrum. However, metal conduction bands are partly filled bands, and any excitation energy will be quickly dissipated through recombination of the generated electron and hole so that metallic systems are not suitable light-harvesting systems.

From the Jablonski diagram of molecular systems (Figure 2.9) we learned that deactivation is fast within the vibrational excited state manifold, but that this process slows down by several orders of magnitude at the vibrational ground state of the LUMO. This allows competing processes like energy transfer and charge separation to play an important role. The situation is similar in semiconductors, where a band gap exists between the filled valence band (VB) and the empty conduction band (CB), just as in molecules between the HOMO and the LUMO. It is necessary to separate the charges and collect them at different electrodes while retaining as much as possible of the band gap energy in the absorber as a useful voltage U in the electrical circuit (see Figure 7.10).

It is important that the electrode materials are chosen such that they match the levels of the semiconductor band to minimise losses. Nevertheless, a small gradient (typically a few tens of a millivolt) needs to remain in order to drive the charge separation. No process occurs unless there is a gradient that provides a driving force.[35] Charge separation can occur by drift of the charge carriers under the influence of an externally applied electric field. Alternatively, it can occur by diffusion from regimes of higher to lower concentrations, following a gradient in the electrochemical potential.

7.4.3 The Shockley–Queisser limit

The photons with an energy too small to overcome the band gap E_g are not absorbed and therefore lost for photovoltaic conversion. Secondly, the relaxation of the excess energy $h\nu - E_g$ of hot charge carriers is lost, mostly as heat. The two effects represent the main contributions to the losses in a single junction cell. These losses were calculated by Shockley and Queisser by integration of the corresponding energy in the solar spectrum as a function of the bandgap energy.[36] Wien's law of black-body radiation predicts the product of the temperature and wavelength of the intensity maximum, $\lambda_{max} T = 2.8986 \times 10^6$ nm K. For a solar surface temperature of $T = 5778$ K this gives the observed $\lambda_{max} = 500$ nm. However, the energy distribution of black-body radiation is non-symmetric, and the maximum theoretical efficiency is obtained at 33.2% for a band gap of 1.34 eV, or 925 nm. The value of 33.2% is called the

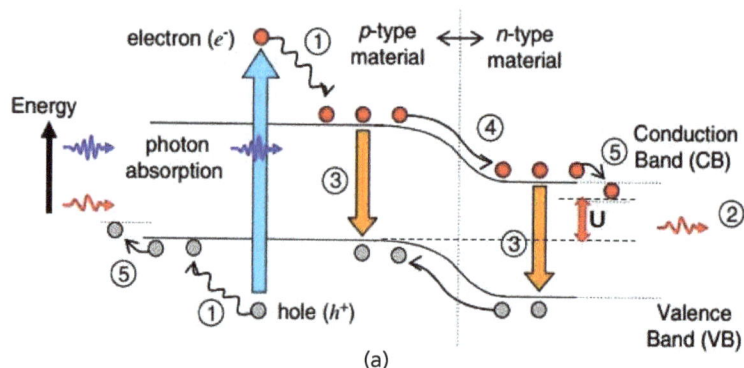

(a)

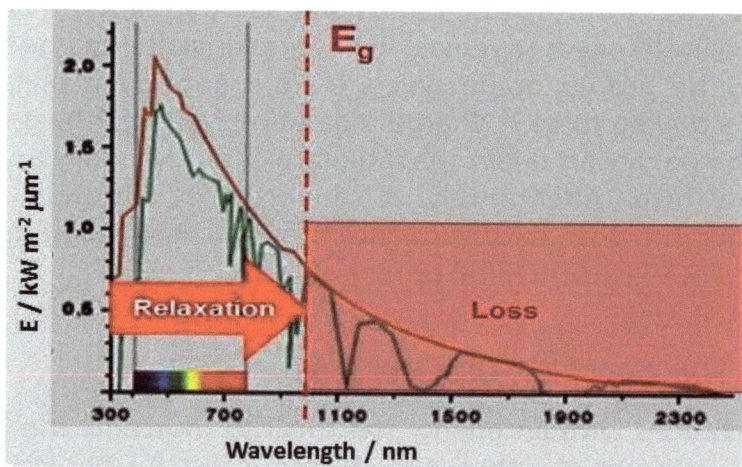

(b)

Figure 7.10: (a) Schematic representation of the semiconductor bands near a p–n junction and the processes occurring after excitation: electron relaxation within the CB and hole relaxation within the VB (1), emission of non-absorbed photons (2), charge recombination (3), and band offset at junction (4) and at electrodes (5). The useful voltage in the electrical circuit is U. (b) Main contributions to the Shockley–Queisser limit in a semiconductor of band gap E_g: photons with $h\nu < E_g$ are not absorbed and therefore lost (2), the excess energy of those with $h\nu - E_g > 0$ is also lost due to relaxation process (1), (reprinted from Ref. [35], open access).

Shockley–Queisser limit (Figure 7.10(b)). This optimum band gap lies between that of Si (1.1 eV) and the one of GaAs (1.43 V). TiO_2 is an abundant and cheap semiconductor and popular support in catalysis and also for solar cells. However, the material is white, meaning that it does not absorb in the visible. It is actually used as a pigment for white suspension paint, and its absorption edge is just below the 400 nm limit of the human eye. Only about 5% of the solar energy has a wavelength that exceeds this band gap. It would therefore not be a good choice for a solar cell harvesting material that works by direct absorption across the band gap. As will be seen further below, sensitisation provides a way to circumvent this difficulty.

Two different efficiencies are quoted to describe the performance of solar cells: (i) the incident photon to current efficiency (IPCE) which measures the ratio of carriers collected at the electrodes to the number of incident photons, and (ii) the light to electric power conversion efficiency (PCE) that gives the fraction of energy that is converted from light to electric power. About 90% of current photovoltaic cells are made from monocrystalline silicon with an energy efficiency of ~25% or from polycrystalline silicon with ~18% PCE. The remaining 10% are thin film cells, consisting mostly of microcrystalline or amorphous silicon, but also copper–indium–gallium–diselenide (CIGS) or copper–indium-disulfde (CIS) with 22% efficiency. Alternatives are CdTe cells with PCE up to 22% for single cells under laboratory conditions but ~10% for modules. Silicon solar cells have lifetimes of 20 years and more with relatively little degradation, but Si is expensive due to the required high purity. Indium is a rare element and thus not suited for applications on a large scale. Cadmium and tellurium are poisonous and should not accumulate in the environment, in particular not in the air as in case of a fire.

7.4.4 *Organic solar cells*

Environmentally friendlier, low cost alternatives are currently being developed. One option that is normally used for direct excitation are organic solar cells (also called plastic solar cells). They are based on conjugated organic polymers which act as absorbers and electron donors (Figure 7.11). The values for the band gaps (E_g) are easily obtained from optical

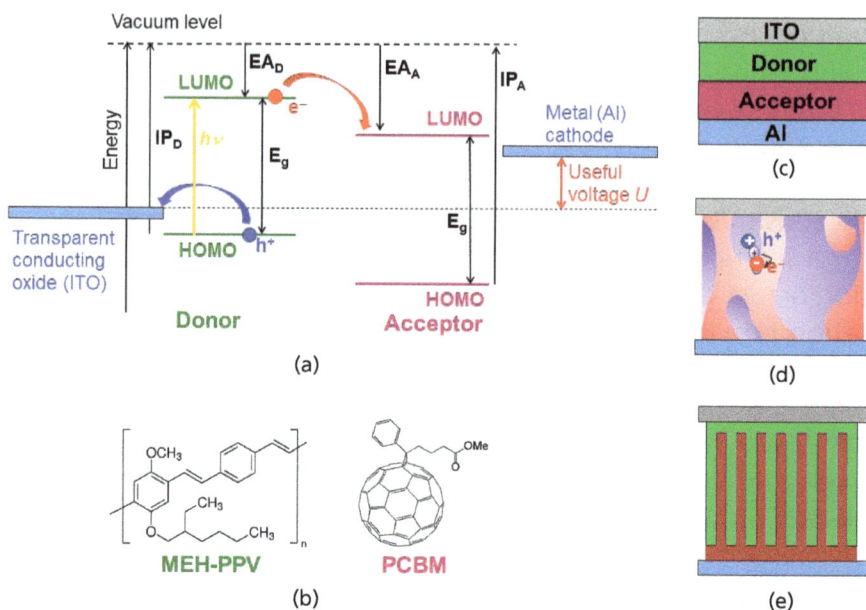

Figure 7.11: (a) Energy-level scheme of a typical two-component organic solar cell with a donor that absorbs the light and transfers the excited electron from its LUMO to the LUMO of an acceptor and finally to the metal cathode, here Al, while the hole remains on the donor and is transferred to the transparent conducting oxide electrode, here indium tin oxide (ITO). (b) Structures of a typical donor polymer, poly[2-methoxy-5-(2-ethylhexyloxy)-1,4-phenylenevinylene] (MEH-PPV), and a common acceptor polymer, [6,6]-phenyl-C_{61}-butyric acid methyl ester (PCBM). (c) Layer structure. (d) Polymer blend. (e) Idealised interdigitated structure with desired 10–15 nm wide pins to optimise charge separation.

spectroscopy, but the absolute values are relevant for the design of high efficiency cells. Absolute values are determined from measurements of ionisation potentials (IP) and electron affinities (EA) or in solution from the half-wave potentials in cyclic voltammetry.[37]

Various conjugated polymers of small molecule monomers (see example in Figure 7.11(b)) are used as absorbers with high extinction coefficients and a band gap of about 2 eV.[38] In principle, cells with a single layer of such a material can produce a photovoltage, but in reality they do not work well because of inefficient charge separation. Blending the acceptor polymer with high electron affinity molecules, typically based on C_{60}, under formation of interpenetrating networks leads to ultra-fast electron transfer at the acceptor–donor interface on a time scale of ~40 fs. Since there are no competing deactivation processes of the excited electron–hole pair (exciton) on this time scale an optimised polymer blend leads to charge dissociation with nearly 100% efficiency.[39] The two polymers can be arranged in a double layer (Figure 7.11(c)), but the thickness of the donor layer needs to be at least 100 nm thick to absorb enough light, much more than the typical diffusion length of excitons that is typically only around 10 nm in these materials. This means that only a small fraction of excitons can reach the heterojunction interface. Dispersed bulk heterojunction systems (Figure 7.11(d)) can reach the 10 nm limit but do not connect the polymers selectively to the proper electrodes. Interdigitated architectures (Figure 7.11(e)) would be ideal but are difficult and expensive to prepare at the required 10 nm scale. More recently, various ternary configurations with two donors and one acceptor or one donor and two acceptors have been investigated and found to achieve power conversion efficiencies up to 11%.[40]

Advantages of organic solar cells include low-cost technology based on printing or roll-on manufacturing processes resulting in low-weight flexible photovoltaic modules for portable or even wearable applications. The choice of solvents and solvent mixtures during the production process is perhaps more an art than a science. Self-assembly processes are also being considered. Morphology is essential, and the obtained heterostructures have to be annealed to minimise traps and increase conductivity. Protective coatings are applied to suppress the fatal access of humidity and oxygen. Nevertheless, stability and lifetime remain an important issue.

The low energy conversion efficiency even under ideal conditions as given by the Shockley–Queisser limit is not satisfactory. It is due to the large losses over the width of the solar spectrum by photons below and above the band gap. Therefore, several alternatives are being considered to reduce losses, as described next.

7.4.5 Multi-junction cells

The first option is based on slicing up the solar spectrum and using each slice in basically separate but combined solar cells, as explained in Figure 7.12. The simplest option consists of two cells and is called a tandem solar cell. The individual cells are usually hooked up in series with two terminals, which requires that each cell converts the same number of photons to electrons per unit time to avoid bottleneck effects. Alternatively,

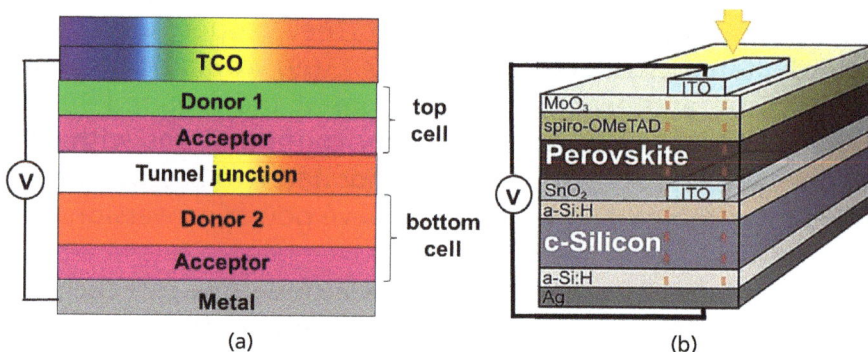

Figure 7.12: (a) Schematic drawing of tandem solar cell. The entire solar spectrum has access through the transparent conducting oxide (TCO) to the top cell with a high band gap donor 1 that absorbs the blue and green part of the spectrum and converts it to a photo-voltage. The yellow and red light is transmitted and enters the bottom cell where it is absorbed by the lower band gap donor 2 and converted to a second photo-voltage. The two cells are separated by a transparent and conductive tunnel junction. When hooked up in series the two photo-voltages add up. For proper function the number of converted photons needs to be the same in both cells, which is adjusted by the thickness of the two donor absorbers. (b) Specific architecture of a tandem cell consisting of a perovskite and a crystalline silicon cell (adapted with permission from Ref. [42]. © (2016). Royal Society of Chemistry).

they can also be connected in parallel with an additional terminal to the intermediate layer. The resulting efficiency depends on the fill factor of the combined device.[41] More straightforward and easy to replace in case of failures is a simple stacking of independent cells that use different absorber materials and have transparent electrodes. The current and voltages are then combined from the entire modules instead of on cell level.

Commercial tandem solar cells with efficiencies of 30% under one sun illumination can reach 40% for concentrated light. However, the complexity of multi-junction cells makes them expensive, which limits their use to selective applications, as in space.

7.4.6 *Upconversion and downconversion*

Since the large losses that lead to the low energy efficiency of solar cells are due to the large width of the solar spectrum it is attractive to think about reducing the width of the spectrum. This can be attempted either by combining two or more low energy photons to a single photon of higher frequency. This has been dubbed upconversion. Alternatively, high energy photons could be "cut into two" to add them to the low frequency part of the spectrum, which is called downconversion.

There are different options to achieve upconversion. The first one is frequency doubling, also called second harmonic generation, which is a nonlinear optical process that occurs in special materials. Effectively, it combines two equal photons to create one photon of double the frequency. This can happen in intense laser fields, and the efficiency increases with the laser power squared, but it is not an option at the intensity of sunlight.

A second option of upconversion involves rare earth ions. Lanthanide ions have a $4f^n5s^25p^6$ electron configuration with $0 \leq n \leq 14$. 14 electrons in 7 orbitals lead to many possible combinations, all of different energies. The 5s and 5p shells shield the 4f inner shell so that the electronic transitions are independent of the surrounding host material.[43] Furthermore, due to spin–orbit coupling in these heavy ions, spin is not a "good quantum number", and the spin selection rules that are quite rigid in light atom (organic) species do not apply in heavy ions. This means that a multitude of transitions are allowed in this ladder of energy levels (Figure 7.13). Absorption from a ground state (G) leads to a relatively long-lived first excited state (E$_1$) which can absorb another pump photon by excited

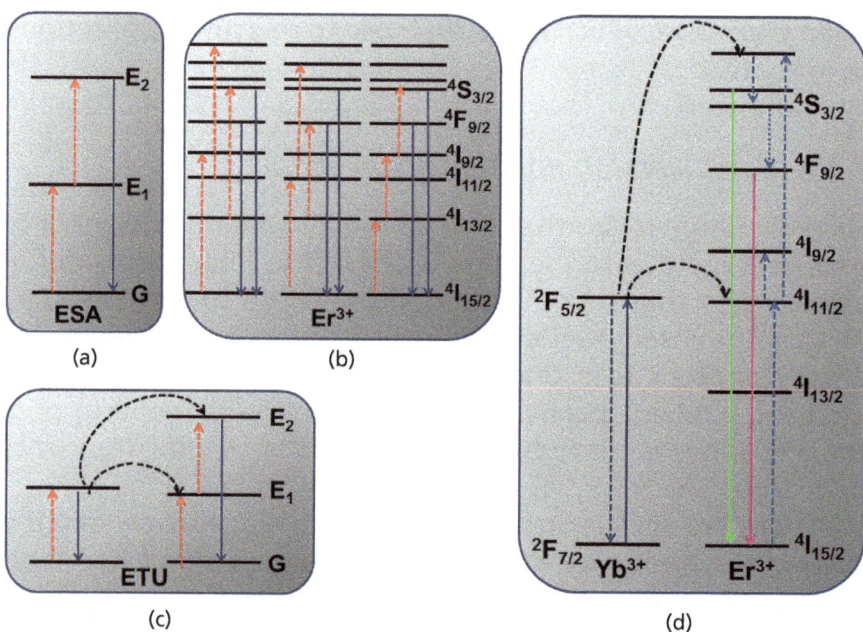

Figure 7.13: (a) Schematic view of upconversion by excited state absorption (ESA) of a single ion. (b) Options for upconversion of low energy photons (red) to higher energy in the rare earth ion Er^{3+}. (c) Schematic view of energy transfer upconversion (ETU) that needs a second ion and (d) ETU from Yb^{3+} to Er^{3+}. In (b) and (d) the states are labelled by the term symbols that are given by a capital letter that describes the total orbital angular momentum of the ion, preceded by a superscript that gives the spin multiplicity and followed by a subscript that gives the quantum number for the total angular momentum derived by vector addition from orbital and spin angular momentum (figure adapted with permission from Ref. [43]. © (2015). Royal Society of Chemistry).

state absorption (ESA) to reach a second and possibly further excited states (E_2) from which the system can deactivate back to G by emission of a photon of higher energy. The most efficient lanthanide-based upconversion mechanisms are two-step ESA of a donor ion followed by energy transfer to a nearby second ion that acts as an acceptor and emitter.[43]

The quantum efficiency of these processes is quite high, but since the 4f shell is shielded and nearly independent of the chemical environment the spectral lines are narrow. Much of the light between the lines is not absorbed, and the energy efficiency for photovoltaic applications is therefore low. Instead, these systems have found interesting applications in solid state lasers, flat screen displays and in particular as bio-labels and for bio-imaging.[43] Compared with organic dyes and QDs, lanthanum-doped nanoparticles have many advantages, including the very low self-fluorescence background and the large Stokes shift that permits a clean separation from photoluminescence or scattering due to the exciting light. They are furthermore inert to photo-bleaching, and excitation in the infrared benefits from the optical transparency window and allows deep tissue penetration.[43]

A convincing option for highly efficient photon upconversion is based on triplet–triplet annihilation following triplet energy migration. It has been realised in a wide range of organic chromophore assemblies in ionic liquids, amorphous solids, gels, supramolecular assemblies, molecular crystals and metal–organic frameworks (MOFs).[44] For a maximum upconversion quantum yield at weak solar irradiance and minimal air sensitivity it is essential to have control over the geometric arrangement, in particular the distance, of the involved pairs of donor and acceptor molecules (Figure 7.14). Liquid solutions cannot compete with solid arrangements since the chromophores are often not sufficiently soluble for the high concentrations needed to provide the short transfer distances.

As seen in Figure 7.14, the mechanism needs two identical donor–acceptor pairs. The donors are excited and chosen such

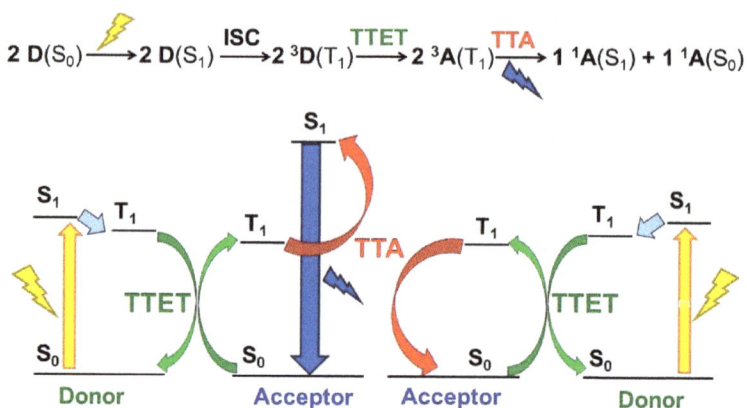

$$2\,D(S_0) \xrightarrow{} 2\,D(S_1) \xrightarrow{ISC} 2\,{}^3D(T_1) \xrightarrow{TTET} 2\,{}^3A(T_1) \xrightarrow{TTA} 1\,{}^1A(S_1) + 1\,{}^1A(S_0)$$

Figure 7.14: Mechanism of upconversion by triplet–triplet annihilation (TTA). Two donor molecules are excited to their S_1 state and undergo inter-system crossing to their longer-lived T_1 states, which then undergo triplet–triplet energy transfer to two acceptor molecules in close proximity. The two acceptor molecules can then undergo a spin-allowed TTA process, during which the electron in one of the two triplet states is promoted to the excited S_1 state, using the energy released from the second acceptor molecule, which decays to the ground state. The first acceptor molecule finally emits the higher energy photon when decaying from S_1 back to the ground state (adapted with permission from Ref. [44]. © (2016). Royal Society of Chemistry).

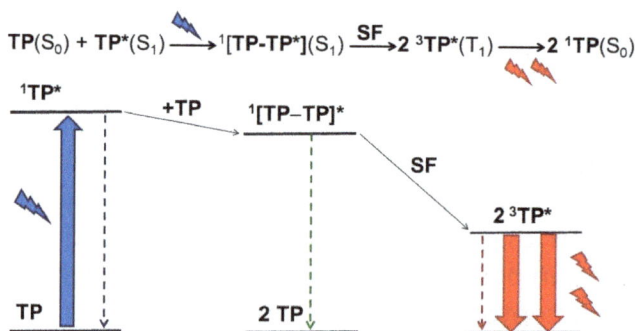

$$TP(S_0) + TP^*(S_1) \xrightarrow{} {}^1[TP\text{-}TP^*](S_1) \xrightarrow{SF} 2\,{}^3TP^*(T_1) \xrightarrow{} 2\,{}^1TP(S_0)$$

Figure 7.15: Mechanism of downconversion by singlet fission (SF). One excited singlet TIPS-pentacene (TP) molecule and one ground state molecule associate in the rate-limiting step to an excimer. The triplet state of TP is low enough so that the excimer can undergo fission to the two monomers in their triplet state, with opposite electron spin. They emit their energy with high efficiency as fluorescence, which provides two low energy photons from one of higher energy (adapted from Ref. [45]).

that they have a large intersystem crossing rate and end up in the longer-lived triplet state. These triplet donor molecules have to transfer their energy to an acceptor that has a slightly lower triplet energy but a much higher excited singlet (S_1) state. When two acceptors in the triplet state are in close proximity the two excitation energies can relocate to a single molecule. If this happens by the Dexter energy transfer mechanism (Section 6.2) that involves an electron transfer and a back-transfer of an electron with opposite spin this leaves both molecules in a singlet state, one in the highly excited S_1 and the other one in S_0. Dexter transfer needs orbital overlap; the two exchanging partners therefore need to be within a distance of about 1 nm or less (Section 6.2). The acceptor emits a single high energy photon that has, depending on the chosen system, nearly double the energy of the two photons absorbed by the donors.

The process can be extremely useful in photovoltaic cells to overcome the Shockley–Queisser limit. It may also be applied to photodynamic therapy of tumours. In this case, biological tissue is not sufficiently transparent to visible light, but near-infrared light at wavelengths ≥ 700 nm is transmitted and converted to more powerful blue light that is absorbed by the tumour and damages its tissue.

Downconversion or quantum cutting is the opposite strategy to upconversion. Its aim is circumventing the Shockley–Queisser limit of photovoltaic cells. The process that "cuts" a high energy photon into two of lower energy in combination with low bandgap solar cells is illustrated in Figure 7.15 for bis(triisopropylsilylethynyl (TIPS)) pentacene (TP). For such single-junction solar cells the theoretical efficiency limit increases to 44%.[45] In such a system, a high energy singlet excited state forms an excited state dimer with an identical molecule in the ground state. Subsequently, this decays to form two triplet monomers with a net spin of zero. Therefore, it conserves spin and is a faster option to produce triplet states than the above described intersystem crossing of organic chromophores. It has been established for solid films of pentacene that this fission process is exergonic. It is rapid and efficient[45] since it makes triplet formation competitive to decay of singlet excited TP* by fluorescence and radiationless decay. The two TIPS substituents lead to favourable intermolecular orientations, high hole mobility in the solid state, and it increases the solubility in chloroform.[45] It has been reported that for solid rubrene the maximum population of correlated triplet states was reached within 20 femtoseconds following singlet excitation.[46] A peak external quantum yield of 109 ± 1% has been reported, and it was suggested that the triplet yield approached 200% for pentacene films thicker than 5 nm.[47]

Triplet states and holes are sensitive to quenching by molecular oxygen and water. A photocurable fluoropolymer layer was applied as a protective surface barrier. In this context, it was discovered that the fluoropolymer had the additional effect to shift the UV part of the solar spectrum into the visible and thus into the regime where perovskite absorbers are active. This represents an additional mechanism that can increase the energy efficiency by narrowing the solar spectrum.[48]

7.4.7 Light harvesting by sensitisation

The principle of dye sensitisation was discovered in 1873 by H. W. Vogel who used it in photographic emulsions to add sensitivity to green, yellow, orange and red colour.[49] The concept became extremely useful in practical applications to solar cells.[50] It is based on 10–20 nm diameter sintered nanoparticles of a semiconductor, mostly TiO_2 in the form of anatase, which provide a large surface that is functionalised by a dye that is grafted via carboxylate groups to the semiconductor surface (Figure 7.16). The excited state D* must be located energetically above the lower band edge of the semiconductor so that it can transfer the excited electron in the HOMO by "injection" into the semiconductor via Dexter energy transfer (Section 6.2). If electron injection is rapid compared to competing deactivation of the dye this provides the first step of charge separation.

The electron then travels from grain to grain, taking advantage of the network of bridges introduced by sintering, and onwards to a conducting transparent oxide electrode (CTO) that transmits the incoming light. Originally, this electrode often consisted of indium tin oxide (ITO), but due to the price of indium that is a relatively rare element, an alternative was needed and found in the form of fluorine doped tin oxide (FTO). In order to minimise voltage losses it is necessary that its Fermi level is just below the lower band edge of nano-size TiO_2.

Since absorption of the dye is in the visible part of the spectrum but that of TiO_2 in the UV just below 400 nm, transfer of the hole is energetically not possible. It has to be cleared with

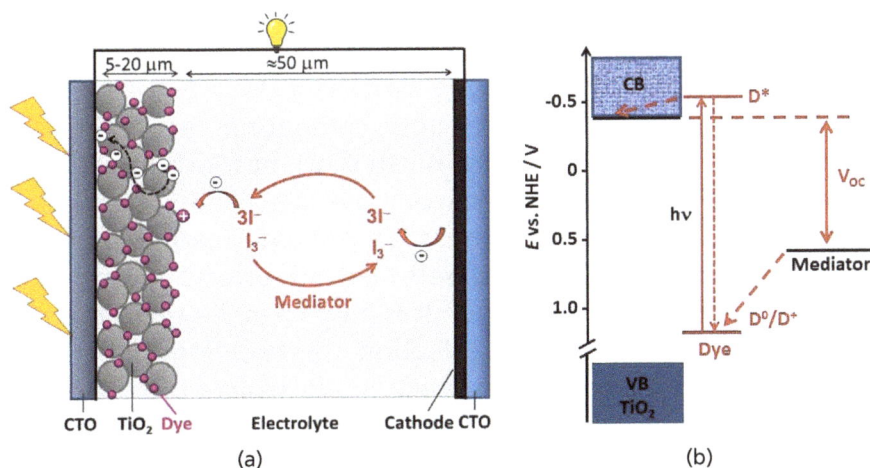

(a)   (b)

Figure 7.16: (a) Schematics of dye sensitised solar cells (Grätzel cell) with organometallic molecular complex dye (purple dots), grafted onto a TiO_2 support, and (b) corresponding energy level scheme. The dye is excited by a photon $h\nu$ in the visible spectral range. Its excited state D* needs to be located energetically above the lower edge of the TiO_2 conduction band, allowing the excited electron to be injected into the conduction band (CB) of TiO_2 and transferred onto a conducting transparent oxide (CTO) anode. The hole on the dye molecule is cleared by a redox mediator system, which acts as a shuttle that transports it through the liquid electrolyte to the cathode. The maximum output voltage is the open circuit voltage V_{OC}, typically about 0.7 V, given approximately by the difference between the level of D* and the potential of the redox mediator.

an electron from the cathode, and this is done by a redox mediator that acts as a shuttle for the electron. A popular mediator is the iodide/triiodide redox couple,

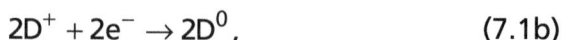

$$3I^- \rightarrow I_3^- + 2e^-, \tag{7.1a}$$

$$2D^+ + 2e^- \rightarrow 2D^0, \tag{7.1b}$$

in which I^- ions donate their electrons to regenerate the oxidised dye under formation of I_3^-, which then diffuses to the back-contact cathode, often Pt, where it is reoxidised and ready for the next cycle. As a dye, Ru(II) complexes displayed satisfactory stability and were the systems of choice for a long time, but promising cheaper alternatives based on a Zn porphyrin dye and a Co(II/III) redox couple have been developed more recently, reaching 12.3% energy efficiency.[51]

As a consequence of the slow competing recombination kinetics, dye sensitised solar cells can work at much lower light intensity than conventional silicon cells, also under cloudy sky and with indirect light. Since heat is more easily carried off from thin structures, this reduces self-heating effects with corresponding decrease in efficiency. It is easy to make tandem cells. Economically, the simple production methodology is a clear advantage. On the contrary, the liquid electrolyte is a disadvantage. It has stability problems and may freeze at temperatures below zero. Cells have to be carefully sealed to avoid evaporation or access of oxygen.

Attempts have been made to replace the liquid electrolyte by a quasi-solid-state gel electrolyte, but a more efficient strategy was to replace the electrolyte by a solid hole conducting material. A dramatic development started in 2009, when Miyasaka et al.[52] used organometal halide *perovskites* as visible light sensitisers on TiO_2 with a liquid electrolyte. Finally, the breakthrough came when Snaith and Lee replaced the electrolyte by a solid state polymer hole conductor.[53] In the following, it was found that the perovskites could conduct holes as well as electrons. This rendered the titania scaffold unnecessary.

By 2016, single junction perovskite solar cells reached a respectable 22% PCE, considering a developing time of only 7 years. The hole transport layer (HTL) can have various compositions, often spiro-type organic polymers containing aromatic groups.[54] A common version of today's architecture of perovskite sensitised solar cells is outlined in Figure 7.17, but other variants show similar PCE values.[54]

The semiconductor band structure is formed by the PbX_6 octahedra, while the organic cation acts more as a spacer between the octahedral units. The cation A is most often methyl ammonium (MA = $CH_3NH_3^+$), formamidinium (FA = $CH_2(NH_3)_2^+$) or Cs^+, B is commonly Pb^{2+}, while X are halide anions, usually iodide or bromide. Mixed perovskite compositions of $FAPbI_3$ and $MAPbBr_3$ showed improved stability and increased energy

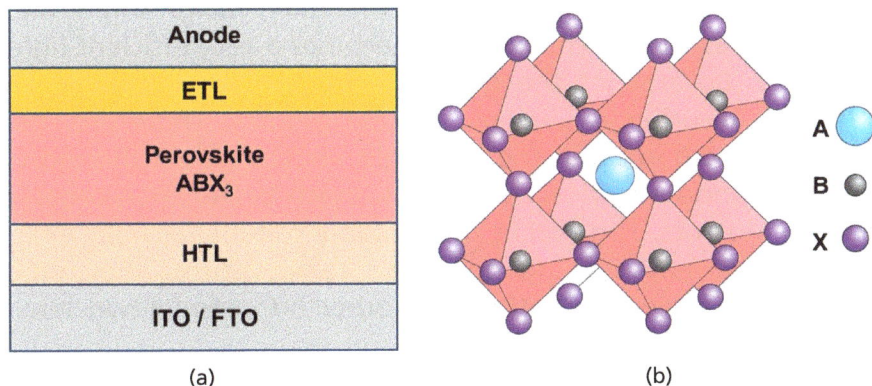

(a)   (b)

Figure 7.17: (a) The all-solid-state version works on the same principle of the dye sensitised cell, but the dye is a perovskite or a semiconductor quantum dot, and the titania scaffold is optional (here omitted). Light absorption in the perovskite dye creates an electron hole pair. The electron transport layer (ETL) consisting often of metal oxides (ZnO, TiO_2, SnO_2) or C_{60} fullerene has a rectifying function as it acts also as a hole blocking layer and transmits only electrons to the anode, while the hole migrates through the hole transport layer to the cathode (Au, Ag), a conducting transparent oxide such as indium tin oxide (ITO) of fluorine doped tin oxide (FTO). (b) Structure of perovskite of general formula ABX_3. A is often the methyl ammonium cation, B in the centre of the octahedron is usually Pb^{2+}, and X is a halide ion.

conversion efficiency. An increased amount of bromide increases the band gap, whereas a higher amount of FA decreases the band gap, but to a lesser extent than bromide.[54] It has proven difficult to replace the toxic Pb^{2+} by a less problematic element, but only Sn^{2+} has shown limited success because of its tendency to oxidise to Sn^{4+}. Lead iodide perovskites show an increase of the band gap upon partial substitution of the larger FA by the smaller Cs, due to tilting of the octahedra. Perovskites based on tin, which is slightly smaller than lead, show a decrease of the band gap due to contraction of the lattice.[55] Absorption is quite broad in these systems — many of them are black — but emission is narrow, with a width of typically $\approx$10–40 nm, and it can be tuned over the entire visible range by changing the composition of the perovskite and also by quantum confinement effects.[56]

Spin–orbit coupling in the caesium lead halide perovskites ($CsPbX_3$, with X = chloride, bromide, iodide) mixes singlet and triplet states so that the material becomes a very efficient light absorber. Because of the quantum-mechanical reciprocity an efficient absorber is also an efficient emitter. Therefore, the best solar cells are also the best light emitters.[57] Furthermore, they are good electrical conductors, much better than organic semiconductors. This makes them highly suitable as light emitting diode materials. At cryogenic temperatures they emit about 1000 times faster than any other hitherto known semiconductor nanocrystal.[58]

One parameter that determines good cell performance is a high open circuit potential V_{OC}. The higher the band gap of the dye, the higher V_{OC}, but partly at cost of other parameters. It is essential to have good band matching with small but non-zero losses in order to maintain transport and suppress back-transport, just as discussed in context with the Shockley–Queisser limit in Figure 7.10 for solar cells with direct band gap excitation. An outstanding value of V_{OC} of 1.19 V has been reached.[54] The second essential parameter is the short-circuit current density, J_{SC}. A high J_{SC} requires complete harvesting of photons over a relatively large spectral range, while maintaining an efficient charge separation, transport and extraction with minimal competing processes such as charge recombination.[54] The Ohmic resistance of each transport process is proportional to the transport length, which should be minimised by nano-structuration. Other resistances occur at each contact between different layers. In context with conductivity it is important to have tempered layer morphologies with a low number of traps.

The long-term stability of perovskite solar cells is still a considerable challenge. The origin of instabilities can mainly be ascribed to unstable perovskite materials, which tend to decompose under environmental stress such as humidity, heat or prolonged light illumination in air.[54] Various methods introducing a protective barrier layer to suppress the permeation of water were developed. The most successful attempt was a coating with a photocurable fluoropolymer that had the additional effect to down-convert the UV part of the spectrum into the visible via fluorescence.[48] This provided encouraging near 19% PCE, and 98% was retained after 6 months.[54] However, Pb is an example of an element of environmental concern that may not be admitted for use in commercial solar cells. If Pb gets banned for solar cells this would constitute a severe obstacle for perovskite cells.

Quantum dot sensitised solar cells are a promising low-cost alternative to existing photovoltaic technologies.[59] The QDs act as a dye with high extinction coefficient for solar light harvesting. The band gap can be tuned in a wide range by tailoring size and composition of the dot, while individual band edges can be influenced by modification of the surface with polar groups.[59] The concept allows various cell architectures, including QD-sensitised wide-bandgap single junction nanostructures that work with liquid electrolytes (in conceptual analogy to the cell shown in Figure 7.11 or fully solid state cells based on QDs blended with polymer hole conductors or QD layers sandwiched between the essential electron and hole conductors to select the separated charge carriers (Figure 7.12(a)).[59] Power conversion efficiencies exceeding 10% have been reached. CdTe has been well-investigated and can absorb multiple frequencies. However, Cd and Te as well as Se in place of Te are rare and highly toxic elements and will have to be avoided in solar cells, considering the contamination of the atmosphere in case that a roof with solar panels catches fire.

7.5 Light Harvesting in Natural Photosynthesis

7.5.1 *Introduction*

Virtually all life on the earth depends on photosynthesis, a process by which solar energy is used to convert carbon dioxide and water into carbohydrates. A wide variety of organisms photosynthesise. Collectively they are responsible for storing energy in the form of biomass at a remarkable average rate of about 130 TW. This is equivalent to 1.5×10^{14} kg of biomass per year.[60] In the process, molecular oxygen is produced by oxygenic photosynthetic organisms at a rate of 4×10^{14} kg per year.[61] Interestingly, oxygen is released as a by-product during water splitting to extract electrons and protons from water. About 50% of the photosynthetic activity takes place in the ocean, mainly by cyanobacteria and diatoms.

Not only does photosynthesis play a pivotal role in the existence or sustenance of all life on earth, but in recent years the photosynthetic process has been considered a source of inspiration for the development of inexpensive, environmentally friendly, renewable solar energy technologies. In particular the first photosynthetic steps — light harvesting and charge separation — are based on remarkable design principles in their use of abundant and low-cost materials to achieve near-unity quantum efficiencies in "noisy" environments. Various bio-inspired organic photosystems have already been engineered; however, their low quantum efficiencies clearly point to the infancy of this new development.[61–65]

Whereas human technologies usually look for optimality, nature often rather targets robustness. This is in particular the case for photosynthesis. Photosynthetic organisms constantly have to adapt to fluctuating environments on all levels of organisation, and this is where they expend most of their energy. Solar technologies based on inexpensive molecular materials, such as organic semiconductors, polymers, and nanoparticles, face similar challenges of noisy environments and phototoxicity and would do well to gain inspiration from the ingenious design of photosynthesis.

7.5.2 *The photosynthetic light-harvesting apparatus*

All photosynthetic organisms use the same blueprint for their light-harvesting apparatus: large networks of light-harvesting antennas are responsible for light absorption and transfer of the resulting electronic excitation energy to the reaction centre, the site where photochemistry takes place (Figure 7.18). The basic photosynthetic unit is called a photosystem and typically consists of a reaction centre connected to a few antenna complexes.[66] Through this design the effective absorption cross-section of the reaction increases with 1–2 orders of magnitude and the dilute energy from the sun is concentrated to such an extent that an optimal photochemical turnover rate is obtained.[67]

Each light-harvesting antenna consists of one or more proteins binding a collection of chromophores in fixed positions and orientations. Despite the presence of usually only one or two types of chromophores in the antenna, a few design principles give rise to the formation of large energy gradients (Figure 7.18(d)). Energy gradients increase the absorption spectral window and create an energy manifold that leads to rapid relaxation and transfer of an excitation to a favourable site within the antenna. The first principle is the use of a protein, a highly heterogeneous dielectric environment. By embedding chromophores at different sites within this environment

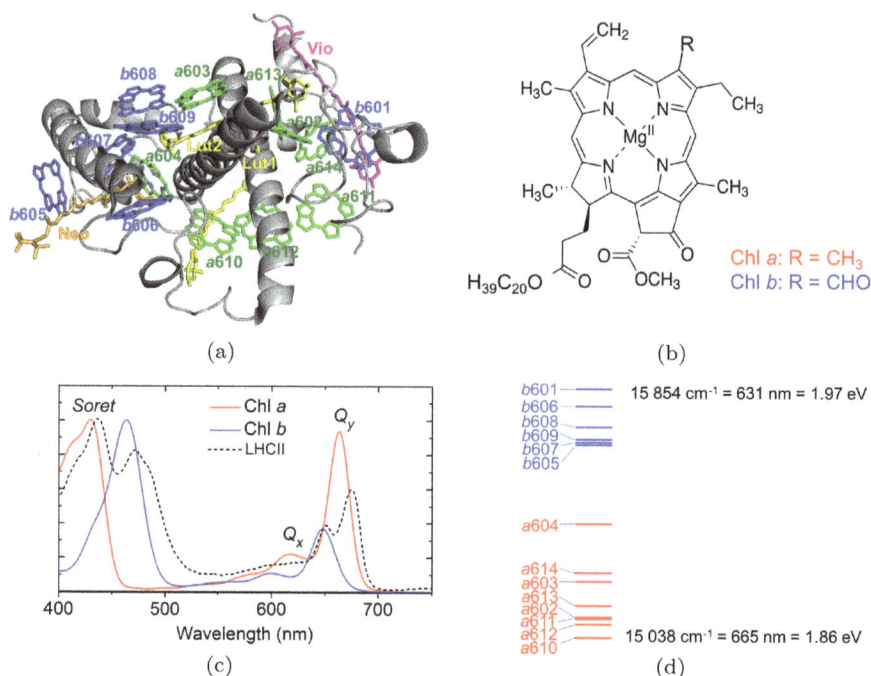

(a)

(b)

(c)

(d)

Figure 7.18: (a) Molecular structure of a monomeric subunit of the major plant light-harvesting complex, LHCII, based on the X-ray crystallographic data of Ref. [68] at a resolution of 2.72 Å, shown from the outside of the membrane (i.e. stromal view), and using the nomenclature of Ref. [67] for all bound chromophores. For clarity, only the chlorin rings of chlorophylls (Chls) *a* and *b* are shown, in green and blue, respectively. The protein is displayed as four interconnected grey ribbons, and the carotenoids lutein, neoxanthin, and violaxanthin are denoted by Lut, Neo, and Vio, respectively. (b) Chemical structure of Chls *a* and *b*. (c) Room-temperature absorption spectrum of Chl *a* (solid line) and Chl *b* (dashed line) in ethanol, as well as LHCII (dotted line). All three spectra are normalised at the Soret peak (an intense band in the blue region of the visible spectrum of chromophores). The three dominant absorption bands, Soret, Q_x and Q_y, are indicated for Chl *a*. (d) Relative values of all Chl site energies in each monomeric subunit of LHCII, according Ref. [69]. The values of the highest and lowest energies are indicated (reprinted with permission from Ref. [70]. © (2016) Elsevier).

provides every chromophore with a unique transition energy. The second principle is the use of exceptionally high chromophore densities. For example, the major light-harvesting complex in plants (Figure 7.18(a)), known as LHCII, has a chlorophyll density of 0.25 M.[68] Some of the chlorophyll pairs in this complex have centre-to-centre separations as short as 8 Å, and the average nearest-neighbour chlorophyll separations are <11 Å,[68] significantly shorter than the van der Waals radius of the chlorophyll ring of ~14 Å and ~16 Å along the Q_x and Q_y molecular axes (Figure 7.18(b)), respectively. Such high chromophore densities not only optimise the antennas' effective absorption cross-sections, but they also give rise to Frenkel exciton states, characterised by strong inter-chromophore couplings, often >100 cm^{-1} (see Section 2.4.8).[69] The electronic properties of the chromophores participating in the exciton states are altered, thus further increasing the absorption spectral window and excitation energy manifold. LHCII employs a third principle by using two slightly different types of chromophores: chlorophyll a and b. These two chlorophylls differ only with respect to the small side chain R in Figure 7.18(c), but this structural change shifts the Q_y transition (i.e. the HOMO–LUMO transition) by 30 nm (0.085 eV) (Figure 7.18(c)).

Energy relaxation amongst exciton states is significantly faster than FRET — see Section 6.3. Furthermore, energy relaxation along an energy gradient does not proceed along all individual exciton states in the manifold but energy levels not shared by the same chromophore can be skipped.[71,72] As a result, exciton states ensure that a significantly smaller number of pathways are explored during excitation energy transfer to the reaction centre, which considerably increase the speed and efficiency. For example, photosystem 1 in plants, which contains ~200 chlorophylls, takes only 20–30 ps to transfer the energy from an absorbed photon to the reaction centre, despite the large size of this supercomplex.[73] This time is >100 times shorter than the excitation decay lifetime of 4 ns of chlorophyll in such a protein environment, which explains the near-unit quantum efficiency of these complexes.

7.5.3 *Photosynthetic energy-transfer models*

Photosynthetic energy transfer is remarkably fast. The process of energy transfer to the reaction centre consists of a large number of steps but is typically completed within tens of picoseconds. The fastest steps are energy delocalisation between strongly coupled chromophores, which takes place within tens to hundreds of femtoseconds. It is not surprising that this information only became available when femtosecond lasers were developed. These processes have been investigated using a variety of advanced ultrafast laser spectroscopy methods (see Section 3.5). However, the details of the energy transfer processes have only started to emerge once sophisticated models were developed. This was not a trivial task, considering that the photosystems are large macromolecular systems, containing large chromophore aggregates, and exhibit a complex interplay between protein and chromophore motions and interactions.

Protein motions can be described by glass-like disorder, which operates on time scales ranging from femtosecond to seconds.[74,75] The fastest motions are collective nuclear vibrational modes, known as phonons, and interact with the electronic excited states of the embedded chromophores. This interaction leads to homogeneous broadening of the absorption and fluorescence spectra and energy shifts due to solvation effects.[76]

A complete picture of energy transfer in these complexes is not possible without the existence of atomic resolution structures of the complexes and the combination of various time-resolved and steady-state spectroscopy experiments and extensive modelling. Furthermore, many complexes are capable of absorbing photons with energies significantly lower than the absorption maximum of the reaction centre, yet they maintain a near-unity quantum efficiency. Stark spectroscopy has shown that efficient light harvesting is possible due to the formation of charge-transfer states that mix into excitonic states.[77]

For a description of the energy transfer processes in the photosynthetic special pair of chromophores we refer to Section 6.4, based on the photosynthetic exciton model presented in Section 2.4.8. In most photosynthetic light-harvesting complexes, the dipolar coupling and site energy difference are comparable, on average, and energy transfer consequently operates between the weak and strong coupling limits (Figure 2.8). In this case, both Redfield and Förster theories fail to provide accurate descriptions and complex combinations of these theories or other, more complex approaches are required.[2]

7.5.4 *Photoprotection*

The photosystems are optimised in a great number of ways. Even slight modifications can dramatically change their light-harvesting functionality, for example a small protein conformational change[78] or a configurational change of a key chromophore.[79–81] The photosystems function optimally under relatively low levels of solar radiation. As a result, the amount of light absorption quickly exceeds the organism's physiological needs during exposure to high irradiation levels. Without proper regulation the highly reactive intermediate products in the photosystems of oxygenic organisms generate harmful reactive oxygen species.

Since sunlight can fluctuate frequently and dramatically in both intensity and spectral quality during the day, mechanisms of energy regulation are required on many different time scales. The capacity for light harvesting is adjustable not only on the macroscopic scale, for example by stomatal responses as well as leaf and chloroplast movements, but also on the molecular scale, where a large number of mechanisms are employed to ensure rapid thermal dissipation of excess excitation energy. In this way, a constant energy throughput is maintained at the photochemical reaction centre, synchronised with its turnover rate, and the accumulation of potentially lethal charge separation products in the reaction centre is avoided.

Photoprotection is a crucial part of photosynthetic light harvesting and consists of a complex set of feedback mechanisms. An important part of this regulation takes place in the antenna network of photosystem 2 of oxygenic organisms, demonstrating the important role of energy loss processes in photosynthesis.[82] Using single molecule spectroscopy techniques (Section 7.1) it has been shown that photoprotection in the antenna systems operates largely by means of fine, photoactive control over the intrinsic protein disorder so that intrinsically available thermal energy processes can be used.[83]

7.6 Solar Water Splitting

7.6.1 Background

With 142 MJ kg^{-1}, hydrogen provides the largest gravimetric energy density of any chemical compound, almost a factor of 3 larger than methane, natural gas or diesel. It is about two orders of magnitude better than that of the best batteries[84] and therefore an attractive medium for energy storage. It can be produced from water and converted back to water with relatively high round-trip efficiency. The water splitting reaction

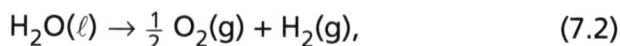

$$H_2O(\ell) \rightarrow \tfrac{1}{2} O_2(g) + H_2(g), \tag{7.2}$$

is strongly endothermic, with a standard reaction enthalpy ΔH_R^0 of +285.8 kJ mol^{-1} and a standard free enthalpy of reaction (or Gibbs Free energy) ΔG_R^0 of +237.1 kJ mol^{-1}. Basically, a single photon of 420 nm has sufficient energy to split one water molecule into its elements, but of course, nobody has ever seen gaseous hydrogen and oxygen bubbling out of water in sunlight. Water is colourless and does not absorb photons from the visible part of the solar spectrum. Also, sensitisation is not straightforward. ΔH_R^0 and ΔG_R^0 relate to the energy differences between reactant and product but do not take into account the transition state of Eq. (7.2) as an activated reaction. Basically, the two O–H bonds of the water molecule have to be broken before new bonds can be formed. This requires 2×458.9 kJ mol^{-1} (which is the average of breaking the HO–H bond at 493.4 kJ mol^{-1} followed by breaking the bond of the O–H radical at 424.4 kJ mol^{-1}). This amounts to 917.8 kJ mol^{-1}, which is much more than ΔH_R^0 of 285.8 kJ mol^{-1}. It is important to realise that the transition state can be lowered when a catalyst is available that stabilises the adsorbed intermediate atoms H* and O* by adsorption free energies $\Delta G(H*)$ and $\Delta G(O*)$. This is shown in Figure 7.19 for the example of the hydrogen evolution reaction

$$2\ H^+ + 2e^- \rightarrow 2\ H* \rightarrow H_2(g). \tag{7.3}$$

For entropy reasons, the adsorption free energy is coverage dependent. In the optimum case it is zero, which means that the H atom binding to the catalyst is about the same as half the bond free energy in H_2. For H on Pt it is straightforward to reach a near-zero adsorption free energy (≈ 0.09 eV or 9 kJ mol^{-1} for a 25% coverage).[86,87] A negative value of $\Delta G(H*)$ indicates a trap state along the reaction coordinate, while a positive value refers to an activated state intermediate. Therefore, any deviations from zero in either the positive or the negative direction will contribute to the activation energy and add to the overpotential of the reaction. This is an essential aspect of the development of co-catalysts which are developed for the hydrogen evolution reaction. For oxygen, the situation is more difficult and further complicated by the fact that the two O atoms in the product molecule stem from two

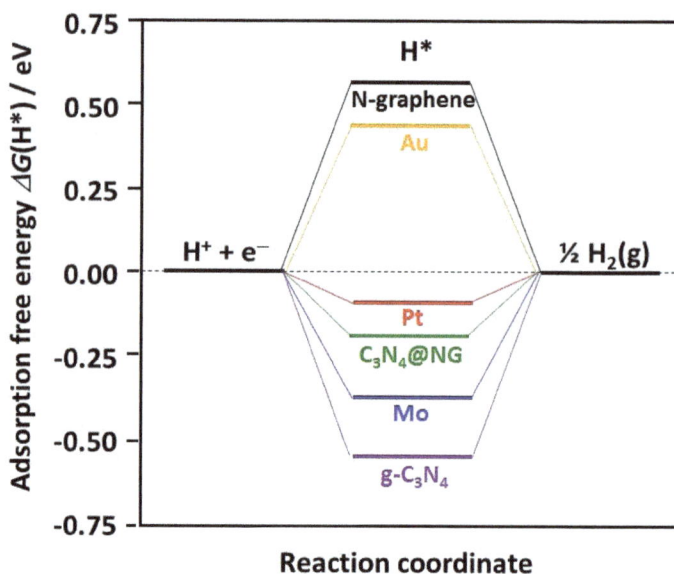

Figure 7.19: Adsorption free energy of hydrogen atoms for an adsorption coverage of 25% on the hydrogen evolution catalysts Pt, Au, Mo, N-graphene (nitrogen doped graphene, NG), graphitic carbon nitride (g-C$_3$N$_4$) and the resulting hybrid structure C$_3$N$_4$@NG. Higher coverages render ΔG(H*) more positive by up to nearly 1 eV, depending on the catalyst (reprinted with permission from Ref. [85]. © (2017) Elsevier).

different water molecules which are not necessarily in close proximity. They have to find each other before recombination can take place.

Water splitting by electrolysis requires the transfer of two electrons at an applied potential of $\geq$1.23 eV. The onset voltage of electrolysis is normally at least 1.5 eV which corresponds to an activation energy of $\geq$0.27 eV (26 kJ mol^{-1}). Most of this activation energy is ascribed to the oxygen evolution reaction.

Commercial electrolysers work at typical current densities of 0.5–2 A cm^{-2}. This is far from what can be reached in photoelectrochemical systems where the limitation is on the order of 10 mA cm^{-2} due to the low flux of the incoming solar radiance and a further reduction by the power-to-current conversion efficiency. This means that to make the same amount of gas a photoelectrochemical solar-to-hydrogen device would require 50–200 times the electrode area of a conventional electrolyser, making the use of precious metal electrocatalysts economically unviable.[88] Instead, one will have to find catalysts which are both very cheap and widely available, and, in addition, one could use concentrators for the incoming solar light.

Water splitting can be carried out in a photoelectrochemical cell where cathode and anode reactions are separated by a proton conducting membrane or diaphragm. It allows hydrogen and oxygen to be collected separately and the progress of the reaction to be monitored by measuring the electrical current. It can also be conducted with the photocatalyst as a powder, deposited in a beaker or dish in the absence of an electrical circuit. Exposure to sunlight leads to the evolution of an explosive mixture of the hydrogen and oxygen gases. For practical use, the two gases will then have to be separated in a second step. The electrochemical cell is therefore the preferred setup, otherwise, the main principles are the same.

Independent of the method, the reaction can be split into a water oxidation and a proton reduction step:

$$\text{Oxidation: } H_2O \rightarrow 2H^+ + 2e^- + \tfrac{1}{2} O_2 \quad E^\circ = -1.23 \text{ V}$$
$$+0.059 \text{ pH vs. NHE,}$$

$$(7.4a)$$

$$\text{Reduction: } 2H^+ + 2e^- \rightarrow H_2 \qquad E^\circ = 0.00 \text{ V}$$
$$-0.059 \text{ pH vs. NHE,}$$

$$(7.4b)$$

$$\text{Overall: } H_2O \rightarrow H_2 + \tfrac{1}{2} O_2 \qquad E^\circ = -1.23 \text{ V.} \qquad (7.4c)$$

The standard potentials hold for pH 1 (1 molar H^+ activity, pH $= -\log a(H^+)$) and standard pressure (1 bar) of the two gases at 25°C as defined for the normal hydrogen electrode (NHE). Since the two reaction steps involve either proton production or proton consumption the electrode potentials become pH-dependent as given by the Nernst equation:

$$E_{red} = E°_{red} - \frac{RT}{nF} \ln \frac{a_{red}}{a_{ox}}$$

$$= E°_{red} - 59 \text{ mV per pH unit.}$$

(7.5)

On changing pH, the two half-cell reduction potentials shift in opposite directions by -59 mV per pH unit, but the cell potential for the overall reaction, which is the sum of oxidation (Eq. (7.4a)) and reduction (Eq. (7.4b)), remains the same. The absolute positions of the band energies of a photocatalyst are essential, independent of whether the reaction is conducted photoelectrochemically or simply photocatalytically. Furthermore, the electrodes (green boxes in Figure 7.20) need to have a Fermi energy that allows the electrons or holes to be transferred. Water splitting is naturally limited to aqueous environments, and the potentials have to straddle the potential of the hydrogen oxidation reaction (upper band edge of valence band) and of the hydrogen reduction reaction (lower band edge of the conduction band), as illustrated in Figure 7.20. The water redox potentials can be shifted to a limited extent by adjusting pH in accord with Eq. (7.5). However, many of these semiconductors are unstable at extreme pH so that a near-neutral pH is preferred.

A general chemical principle states that molecules with a large HOMO–LUMO gap and compounds with a large band gap are more stable than those with a small gap. This is on the ground that large band-gap compounds tend to have a large ionisation potential and a small electron activity and therefore a low driving force in particular for redox reactions. Since the

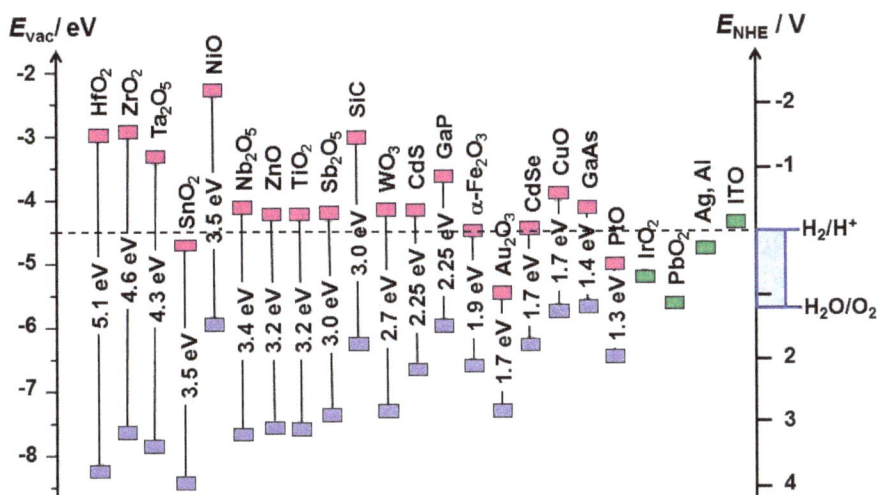

Figure 7.20: Absolute positions of the valence band of various intrinsic semiconductors (upper edge of blue boxes) and of conduction bands (lower edge of red boxes) and band gaps in the order of decreasing values. Metallic conductors are represented by green boxes with the upper edge giving the work function (ITO: indium tin oxide). The band positions are given against the ionisation limit in vacuum (left) and for aqueous electrolytes against the normal hydrogen electrode (NHE, right). Also, shown are the relevant potentials for the water splitting reaction (based on values taken from Refs. [89, 90]).

fraction of solar photons in the UV, with $\lambda < 400$ nm, is rather small, the band gap should be below 3.2 eV (Figure 7.9). This limits severely the choice of semiconductors for direct excitation across the band gap, although they may still be suitable as a scaffold for excitation by sensitisation, in close analogy with the principles of dye sensitised solar cells.

Many of the semiconductor examples in Figure 7.20 are not suitable for direct excitation because their band gaps are too high. Furthermore, the absolute position of the relevant bands of Au_2O_3 and PtO is too low, but GaAs, CuO, CdSe and α-Fe_2O_3 look promising from the viewpoint of band position.

7.6.2 *Non-electrochemical photocatalytic water splitting*

Figure 7.21(a) displays an example that is suited to highlight how a photocatalytic system works. It is based on a RuO_2 nanoparticle with GaN and ZnO in solid solution, but other systems stable against photocorrosion and having a band gap of ≈ 2 eV (absorption edge near 600 nm) like $LaTiO_2N$, Ta_3N_5 and $Sm_2Ti_2S_2O_5$ work in an analogous manner.[91] The valence band needs to be located below the potential for water oxidation (what is normally quoted is the reduction potential, that has the opposite sign, +0.82 V at pH 7, Eq. (7.4a)), the conduction band above the potential for H^+ reduction (−0.41 eV at pH 7, Eq. (7.4b)). Following excitation, the two charge carriers, electron and hole, migrate and at the surface may react in the desired way with water. However, in the absence of co-catalysts this reaction is activated, slow and in severe competition with recombination, which minimises the efficiency of water splitting.

The presence of a Pt nanoparticle as a co-catalyst in good contact with the photocatalyst particle provides lower energy sites than those on the surface of the photocatalyst. Pt or Rh attract H atoms and are excellent promotors of their recombination to H_2, in accord with the hydrogen evolution mechanism described in Figure 7.19. However, they promote also the backward reaction of water formation from H_2 and O_2 and are therefore of limited value. Transition metal oxides which do not promote water formation from the elements are therefore applied as co-catalysts. NiO_x, RuO_2 and more recently synthesised core-shell nanoparticles with a noble metal or metal oxide core and Cr_2O_3 shell were found to be particularly effective. It was concluded that the chromia shell was transparent to protons, and photo-reduction occurred at the noble metal interface to chromia, while the chromia shell was mainly responsible for the inhibition of the reduction of oxygen to water.[92] It is well known that a Mn complex is the O_2 evolution centre in photosynthesis of green plants, algae and cyanobacteria, and in analogy with this it was found that Mn_3O_4 co-catalyst nanoparticles acted as oxygen evolution sites in solar water splitting.[91] Its function is to provide trapping sites for the holes, promote charge separation and lower the activation energy of the reaction (Figure 7.21(b)), which attenuates the reaction. It is therefore considered to be the greatest challenge for water splitting.[93]

The mechanism of catalytic oxygen evolution is complex and involves the adsorbed intermediates HO* that is oxidised to O* which is then attacked by H_2O under oxidation to form HOO*.[94] Omitting the catalyst, it can formally be written as a 4-electron oxidation process of two water molecules by 4 photo-generated holes (h^+):

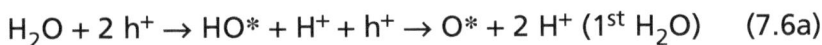

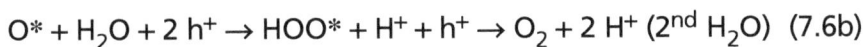

$$H_2O + 2\ h^+ \rightarrow HO* + H^+ + h^+ \rightarrow O* + 2\ H^+\ (1^{st}\ H_2O) \quad (7.6a)$$

$$O* + H_2O + 2\ h^+ \rightarrow HOO* + H^+ + h^+ \rightarrow O_2 + 2\ H^+\ (2^{nd}\ H_2O) \quad (7.6b)$$

The apparent activation energy for the overall water splitting reaction was nevertheless found to be as low as 8 kJ mol^{-1} in the system of Figure 7.21.[91]

Mn_3O_4 has a spinel structure with the oxide ions forming a close-packed cubic structure and Mn(II) occupying tetrahedral and Mn(III) octahedral sites. The structure reminds of the Mn_4CaO_5 water oxidation complex (WOC) that is thought to be responsible for oxygen evolution in photosystem II via a mechanism known as the Kok cycle (Figure 7.22).[95] A similar ligated Mn_4O_4 cube embedded in Nafion was also demonstrated to be amongst the most active and durable molecular water oxidation photo-catalysts.[96] The mixed valence of the Mn ions provides essential flexibility for the redox-catalytic transformations. The water molecules can bind to surface Mn ions and complete their coordination sphere. In this way they become readily available for reaction, and they can restore the catalytic unit after O_2 elimination. In general, the similarity with the Kok cycle is exciting in view of its biomimetic aspect.

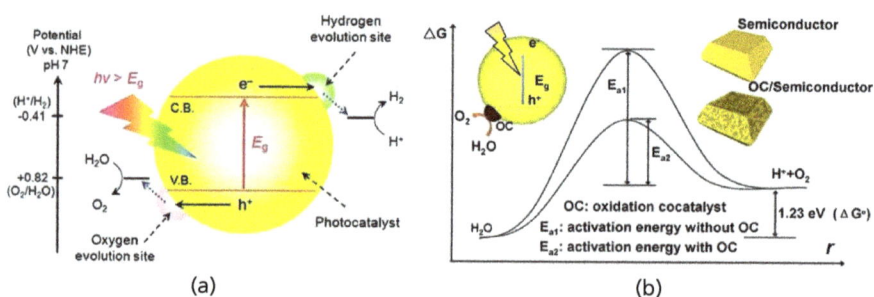

(a)   (b)

Figure 7.21: (a) Schematic energy diagram with valence band (V.B.) and conduction band (C.B.) for photocatalytic water splitting in a single step and in two steps. Loading nanoparticles with Rh or Pt cores and Cr_2O_3 shells as co-catalysts for hydrogen evolution onto a photocatalyst and Mn_3O_4 for oxygen evolution significantly improves the water splitting rate (reprinted with permission from Ref. [92]). © (2014) American Chemical Society. (b) Role of oxidation co-catalyst in photocatalytic water oxidation (reprinted with permission from Ref. [93]. © (2013) American Chemical Society).

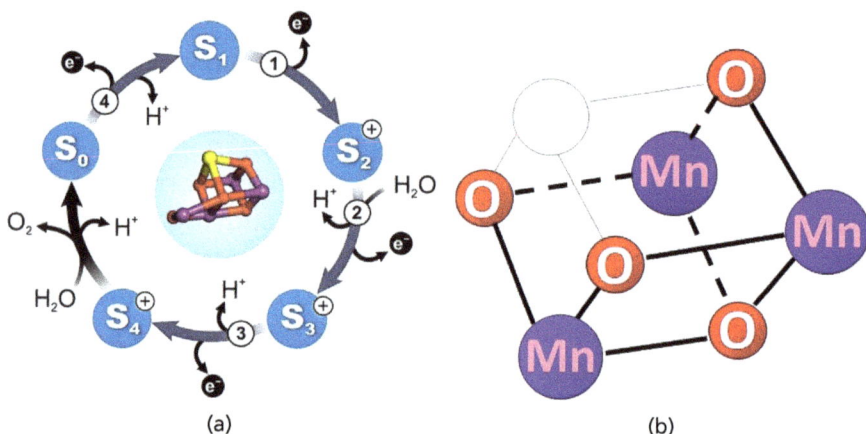

(a)   (b)

Figure 7.22: (a) Kok cycle of oxygen evolution reaction at the water oxidation complex (WOC = Mn_4CaO_5) during water splitting in photosystem II, oxidising two water molecules into molecular O_2 and four proton–electron pairs (reprinted with permission from Ref. [95]. © (2013) Elsevier). (b) Oxygen evolution co-catalyst Mn_3O_4 has a spinel structure where the oxide ions are cubic close packed and the mixed valence Mn(II)/Mn(III) ions occupying tetrahedral and octahedral sites. The Mn_3O_4 unit is a cube with one unoccupied corner (missing Mn, empty circle) that reminds of the WOC in photosystem II where Ca^{2+} occupies the empty corner (yellow atom) and the additional MnO fragment is attached as an appendix to the cube.

In place of the single nanoparticle for direct excitation across the band gap one may design hybrid systems where two different nanoparticles are in contact, one of them acting as light-harvesting system that acts as a sensitiser and injects the excited electron into the second nanoparticle, just as in dye-sensitised solar cells. The light harvesting particle needs to be equipped with the oxygen evolution catalyst, while the second system has to do the hydrogen evolution.[97] The relevant band edges still have to straddle the water oxidation and the hydrogen evolution potentials, as in photovoltaic systems.

7.6.3 *Photoelectrocatalytic water splitting*

Most of the principles derived for non-electrochemical systems hold also for photoelectrocatalytic cells. The difference is that oxygen and hydrogen evolution are spatially separated by a proton conducting membrane (although it also works without membrane but it does not serve the purpose of separate collection of the two gases). The oxygen evolution side contains the light-harvesting system with the proper co-catalyst, and only this side is exposed to light, while the hydrogen evolution normally occurs at a Pt electrode on the dark side. All four photons that are needed for the conversion of two water molecules to one O_2 molecule are spent on the same side. This is the same as in natural photosynthesis where the energy is needed for the water oxidation while the carbon dioxide reduction comes for free. The electrical circuit provides a convenient means for passive monitoring and active analysis of the reaction, e.g. via cyclic voltammetry or impedance spectroscopy, but it is not thought to provide electrical work.

7.7 Applications of Defects in Diamond

Due to their inertness, biocompatibility and exceptional photostability, fluorescent diamond-like carbon nanoparticles are of considerable interest in photonics and as labels in biophysics, materials science, and nanomedicine.[98] The most prominent of the luminescent centres is the nitrogen-vacancy (NV) centre due to its facile formation upon electron irradiation in diamond with abundant substitutional nitrogen atoms (see Section 2.10). Using the methodology developed for single molecule fluorescence spectroscopy (Section 7.1) individual nanodiamonds containing a single NV centre can be studied selectively. An example is presented in Figure 7.23, with part (a) showing a topography image of diamond dispersed on a glass support. It demonstrates a mean crystallite size of 7.5–8 nm. Part (b)

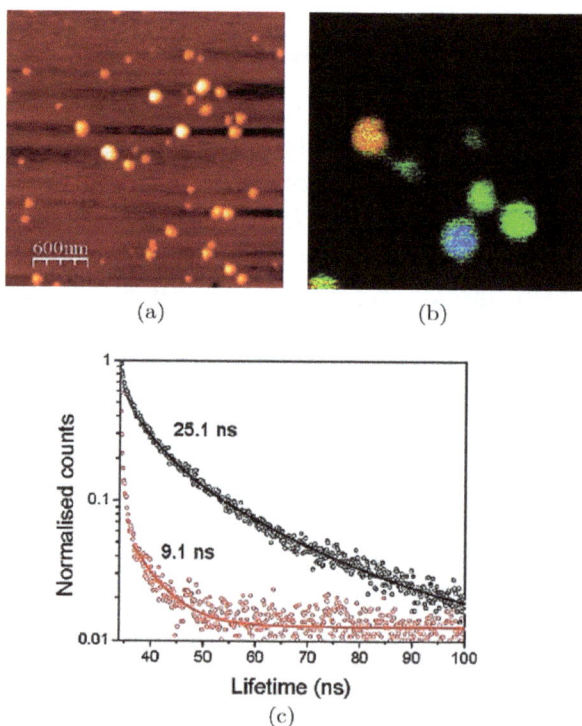

Figure 7.23: Nanosize irradiated diamond containing NV centres. (a) AFM image showing topography. (b) Lifetime coded fluorescence scan of the section shown in (a). Excitation at 532 nm, detection near 700 nm. Colour code: blue: 10 ns, green: 15 ns, yellow: 25 ns. (c) Typical fluorescence lifetime curve of two different nanodiamonds (reprinted with permission from Ref. [98]. © (2009) American Chemical Society).

displays a fluorescence image of the same section. 35% of the diamonds contain a fluorescing defect centre that is detected near a wavelength of 700 nm. The colour in the image represents the fluorescence lifetime which varies between 10 and 25 ns for the different diamonds. These differences were attributed to the varying orientation of the NV electric dipole moment with respect to the glass support.[98]

A more prominent application of NV centres in diamond involves the spin of this paramagnetic defect in view of future quantum information processing.[99] Excitation and fluorescence readout are as described above, but in the dark time between the two optical events the electron spin is manipulated by a radiofrequency pulse that rotates the spin. This method is called optically detected magnetic resonance (ODMR). Coupling of the electron spin to the bath of $I = 1$ nuclear spins of nitrogen, mainly to its large quadrupole moment, causes dephasing of the coherently prepared electron spin precession. A long coherence time is crucial for the storage of information. In a magnetic field of 740 G a coherence time of 3.2 μs was attained for the system of Figure 7.23, which is a significant advantage of nanodiamonds. The decoherence was attributed mainly to spins of the surface termination.[98] For isotopically depleted ^{12}C diamond values up to 2 ms was reported.[100]

7.8 Key Points

- Main applications of optical spectroscopy are in the fields of light harvesting for solar energy conversion (photovoltaic and water splitting), fluorescence microscopy with detection volumes down to 1 femtolitre, time-resolved investigations in natural photosynthesis and in other biological environments, and studies of macromolecular structure and dynamics.
- The phenomena of light absorption, with competing radiationless and radiative deactivation and energy transfer processes are fundamental in all types of applications.

- Intermittent fluorescence (blinking) occurs for all types of fluorophores. It indicates that the system can convert reversibly to dark states (often neutral radicals, radical ions or triplet states), which takes them out of efficient absorption-fluorescence cycles and reduces fluorescence intensity.
- A major application of the various fluorophores (organic dyes, semiconductor quantum dots, or small diamond crystallites containing vacancies neighbouring nitrogen substituted sites) consists in labelling of biomolecules to study their structure and dynamics, often using distance measurements based on FRET. The diamond nanocrystals appear ideal for such applications, while organic dyes may not be sufficiently stable on exposure to light, and the QDs may be too large for accurate distance measurements.
- Energy applications rely on quantitative absorption of solar radiation in a visible to infrared spectral window. Proper alignment of conduction bands of nearby acceptors and valence bands of donors steers charge separation, which is prerequisite for the generation of photovoltaic current.
- Conventional photovoltaic systems consist of silicon or other inorganic solid-state semiconductors. However, in recent years, cheap alternative organic, inorganic or hybrid semiconductor materials have been developed. They are based on direct excitation across the band gap or on sensitisation using an efficient absorber. Their energy efficiencies reach that of silicon cells, in particular for the perovskite absorbers, but their long-term stability towards oxygen or humidity needs further attention.
- Since water is transparent to visible light, water splitting in this spectral range needs sensitisation by an absorber with band gap edges straddling the potentials of the water oxidation and hydrogen reduction potentials. The chemical stability of such low band gap absorbers is often a limiting factor.
- Owing to loss processes of photons with energies higher or less than the band gap of the absorber the maximum

energy conversion efficiency of a single junction photovoltaic or water splitting system is typically around 30%, as described by the Shockley–Queisser limit.

- Charge separation is fundamental also in natural photosynthesis and leads to water splitting following light absorption.

General Reading

- W. E. Moerner, Y. Shechtman, Q. Wang, Single-molecule spectroscopy and imaging over the decades, *Faraday Discuss.*, 2015, 184, 9–36.
- W. E. Moerner, A Dozen Years of single-molecule spectroscopy in physics, chemistry and biophysics, *J. Phys. Chem. B*, 2002, 106, 910–927.
- P. Schwille, E. Haustein, *Fluorescence Correlation Spectroscopy — An Introduction to its Concepts and Applications*, A Tutorial for the Biophysics Textbook online, 2002. www.researchgate.net/publication/235410358 (downloaded 01.03.2018).
- S. Weiss, Measuring conformational dynamics of biomolecules using single molecule fluorescence spectroscopy. *Nature Struct. Biol.*, 2000, 7, 724–729.
- O. Shimomura, Discovery of green fluorescent protein (GFP), (Nobel Lecture), *Angew. Chem. Int. Ed.*, 2009, 48, 5590–5602.
- M. Chalfie, Lighting up life (Nobel Lecture), *Angew. Chem. Int. Ed.*, 2009, 48, 5603–5611.
- R. Y. Tsien, Constructing and exploiting the fluorescent protein paintbox (Nobel lecture), *Angew. Chem. Int. Ed.*, 2009, 48, 5612–5626.
- C. Zander, J. Enderlein, R. A. Keller, *Single Molecule Detection in Solution: Methods and Applications*. Wiley-VCH, New York, 2002.
- S. Rühle, M. Shalom, A. Zaban, Quantum dot solar cells, *ChemPhysChem.* 2010, 11, 2290–2304.
- A. Fontes, R. Bezerra de Lira, M. A. Barreto Lopes Seabra, T. Gomes da Silva, A. Gomes de Castro Neto, B. Saegesser Santos, Quantum dots in biomedical research, *INTECH*, 2012, 12, 269–290. http://dx.doi.org/10.5772/50214.

- A. Mishra, M. K. R. Fischer, P. Bäuerle, Metal-free organic dyes for dye-sensitized solar cells: Property relationships to design rules, *Angew. Chem. Int. Ed.*, 2009, 48, 2474–2499.
- A. Mishra, P. Bäuerle, Small molecule organic semiconductors on the move: Promises for future solar energy technology, *Angew. Chem. Int. Ed.*, 2012, 51, 2020–2067.
- T. Ameri, G. Dennler, C. Lungenschmied, C. J. Brabec, Organic tandem solar cells: A review, *Energy Environ. Sci.*, 2009, 2, 347–363.
- B. C. Thompson, J. M. J. Fréchet, Polymer-fullerene composite solar cells, *Angew. Chem. Int. Ed.*, 2008, 47, 58–77.
- T. P. J. Krüger, V. I. Novoderezhkin, E. Romero, R. van Grondelle, Photosynthetic energy transfer and charge separation in higher plants. In: *The Biophysics of Photosynthesis*, Vol 11, pp. 79–118, J. Golbeck and A. van der Est(eds.); (Series: "Biophysics for the Life Sciences"), Springer, Dordrecht, 2014.
- V. I. Novoderezhkin, R. van Grondelle, Physical origins and models of energy transfer in photosynthetic light-harvesting. *Phys. Chem. Chem. Phys.*, 2010, 12, 7352–7365.
- T. Renger, V. May, O. Kuhn, *Phys. Rep.*, 2001, 343, 137–254.
- H. van Amerongen, L. Valkunas, R. van Grondelle, *Photosynthetic excitons*. Singapore, World Scientific Publishing, 2000.
- R. van Grondelle, B. Gobets, Transfer and trapping of excitations in plant photosystems. In: *Chlorophyll a Fluorescence, A Signature of Photosynthesis*, Papageorgiou CC, Govindjee (eds.), Heidelberg, Springer; 2005. pp. 107–132.
- *Non-Photochemical Quenching and Energy Dissipation in Plants, Algae and Cyanobacteria*, B. Demmig-Adams, G. Garab, W. Adams III, and Govindjee, (eds.), ("Advances in Photosynthesis and Respiration"; Series Editors: Govindjee and T. D. Sharkey), Springer, Dordrecht, 2014.
- P. Muller, X. P. Li, K. K. Niyogi. Non-photochemical quenching. A response to excess light energy. *Plant Physiol.*, 2001, 125, 1558–1566.
- A. Derks, K. Schaven, D. Bruce. Diverse mechanisms for photoprotection in photosynthesis. Dynamic regulation of photosystem II excitation in response to rapid environmental change, *Biochim. Biophys. Acta (BBA) — Bioenergetics*, 2015, 1847, 468–485. https://doi.org/10.1016/j.bbabio.2015.02.008.

References

1. W. E. Moerner, L. Kador, *Phys. Rev. Lett.*, 1989, 62, 2535–2538.
2. T. P. J. Krüger, *From Disorder to Order: The Functional Flexibility of Single Plant Light-Harvesting Complexes.* Doctoral Thesis, Vrije Universiteit Amsterdam, 2011. ISBN: 978-90-8570-766-0.
3. P. Schwille, E. Haustein, *Fluorescence Correlation Spectroscopy — An Introduction to its Concepts and Applications*, A Tutorial for the Biophysics Textbook online, 2002. www.researchgate.net/publication/235410358 (downloaded 01.03.2018).
4. M. Ghafoor, *Synthesis of high refractive index materials for manufacturing apochromatic lens by 3D printing*, MSc Thesis, University of Eastern Finland, 2017.
5. P. Holzmeister, G. P. Acuna, D. Grohmann, P. Tinnefeld, *Chem. Soc. Rev.*, 2014, 43, 1014–1028.
6. T. K. Ha, P. Tinnefeld, *Annu. Rev. Phys. Chem.*, 2012, 63, 595–617.
7. M. Kuno, D. P. Fromm, H. F. Hamann, A. Gallagher, D. J. Nesbit, *J. Chem. Phys.*, 2000, 112, 3117–3120.
8. M. S. Gwizdala, R. Berera, D. Kirilovsky, R. van Grondelle, T. P. J. Krüger, *J. Amer. Chem. Soc.*, 2016, 138, 11616–11622.
9. T. P. J. Krüger, C. Ilioaia, M. P. Johnson, A. V. Ruban, E. Papagiannakis, P. Horton, R. van Grondelle, *Biophys. J.*, 2012, 102, 2669–2676.
10. C. Galland, Y. Ghosh, A. Steinbrück, M. Sykora, J. A. Hollingsworth, V. I. Klimov, H. Htoon, *Nat.*, 2011, 479, 203–207.
11. A. Ruth, M. Hayashi, P. Zapol, J. Si, M. P. McDonald, Y. V. Morozov, M. Kuno, B. Janko, *Nat. Commun.*, 2016, 8, 14521.
12. S. K. Zareh, M. C. DeSantis, J. M. Kessler, J.-L. Li, Y. M. Wang, *Biophys. J.*, 2012, 102, 1685–1691.
13. L. S. Chruchman, Z. Ökten, R. S. Rock, J. F. Dawson, J. A. Spudich, *Proc. Natl. Acad. Sci. USA*, 2005, 102, 1419–1423.
14. K. Paeng, L. J. Kaufmann, *Chem. Soc. Rev.*, 2014, 43, 977–989.
15. K. P. F. Janssen, G. De Cremer, R. K. Neely, A. V. Kubarev, J. Van Loon, J. A. Martens, D. E. De Vos, M. B. Roeffaers, J. Hofkens, *Chem. Soc. Rev.*, 2014, 43, 990–1006.
16. T. Tachikawa, S. Yamashita, T. Majima, *J. Amer. Chem. Soc.*, 2011, 133, 7197–7204.

17. W. E. Moerner, Y. Shechtman, Q. Wang, *Faraday Discuss.*, 2015, 184, 9–36.
18. K. Nienhaus, G. U. Nienhaus, *Chem. Soc. Rev.*, 2014, 43, 1088–1106.
19. M. Strianese, M. Staiano, G. Ruggiero, T. Labella, C. Pellecchia, S. D'Aura, *Fluorescence-Based Biosensors*, In: Spectroscopic Methods of Analysis: Methods and Protocols, Methods in Molecular Biology, Chapter 9, W. M. Bujalowski (ed.), Springer, New York, 2012.
20. https://www.sigmaaldrich.com/life-science/biochemicals/biochemical-products.html?TablePage=16187968.
21. https://www.thermofisher.com/ch/en/home/about-us/partnering-licensing/license-our-technology/intellectual-property-licensing/fluorescent-labelling-detection-technology.html.
22. https://www.thermofisher.com/ch/en/home/life-science/cell-analysis/cell-analysis-learning-center/molecular-probes-school-of-fluorescence/imaging-basics/labelling-your-samples/different-ways-to-add-fluorescent-labels.html.
23. http://www.thermofisher.com/ch/en/home/life-science/antibodies/primary-antibodies/epitope-tag-antibodies/flag-tag-antibodies.html.
24. http://www.fpvis.org/FP.html.
25. U. Resch-Genger, M. Grabolle, S. Cavalliere-Jaricot, R. Nitschke, T. Nann, *Nat. Methods*, 2008, 5, 763–775.
26. S. Weiss, *Nat. Struct. Biol.*, 2000, 7, 724–729.
27. A. Miyawaki J. Llopis, R. Heim, J. M. McCafferi, J. A. Adams, M. Ikura, R. Y. Tsien, *Nature*, 1997, 388, 882–887.
28. H. Noji, R. Yasuda, M. Yoshida, K. Kinosita Jr., *Nature*, 1997, 386, 299–302.
29. H. Sielaff, M. Börsch, *Phil. Trans. R. Soc. B*, 2012, 368, 20120024.
30. W. Kong, X. Yang, M. Yang, H. Zhou, Z. Ouyang, M. and Zhao, *Trends Analyt. Chem.*, 2016, 78, 36–47.
31. W. R. Seitz, in: *Treatise on Analytical Chemistry*, 2nd Ed., P. J. Elving, E. J. Meehan, I. M. Kolthoff (eds.), Part I, Vol. 7, Wiley, New York, 1981, p. 169.
32. O. Adegoke, P. B. C. Forbes, *Talanta*, 2016, 146, 780–788.
33. C. Han, H. Li, *Analyt. Bioanalyt. Chem.*, 2010, 397, 1437–1444.
34. I. Costas-Mora, V. Romero, I. Lavilla, C. Bendicho, *Trends Analyt. Chem.*, 2014, 57, 64–72.

35. E. Roduner, S. G. Radhakrishnan, *Chem. Soc. Rev.*, 2016, 45, 2768–2784.
36. W. Shockley, H. J. Qeisser, *J. Appl. Phys.*, 1961, 32, 510–519.
37. J. Heinze, *Angew. Chem. Int. Ed.*, 1984, 23, 831–847.
38. Ch. J. Brabec, N. S. Sariciftci, J. C. Hummelen, *Adv. Funct. Mater.*, 2001, 11, 15–26.
39. M. C. Scharber, D. Mühlbacher, M. Koppe, P. Denk, Ch. Waldauf, A. J. Heeger, Ch. Brabec, *Adv. Mater.*, 2006, 18, 789–794.
40. Q. An, F. Zhang, J. Zhang, W. Tang, Z. Deng, B. Hu, *Energy Environ. Sci.*, 2016, 9, 281–322.
41. T. Ameri, G. Dennler, Ch. Lungenschmied, Ch. J. Brabec, *Energy Environ. Sci.*, 2009, 2, 347–363.
42. S. Albrecht, M. Saliba, J. P. Correa Baena, F. Lang, L. Kegelmann, M. Mews, L. Steier, A. Abate, J. Rappich, L. Korte, R. Schlatman, M. K. Nazeeruddin, A. Hagfeldt, M. Grätzel, B. Rech, *Energy Environ. Sci.*, 2016, 9, 81–88.
43. X. Li, F. Zhang, D. Zhao, *Chem. Soc. Rev.*, 2015, 44, 1346–1378.
44. N. Yanai, N. Kimizuka, *Chem. Commun.*, 2016, 52, 5354–5370.
45. B. J. Walker, A. J. Musser, D. Beljonne, R. H. Friend, *Nat. Chem.*, 2013, 5, 1019–1024.
46. I. Breen, R. Templaar, L. A. Bizimana, B. Kloss, D. R. Reichman, D. B. Turner, *J. Amer. Chem. Soc.*, 2017, 39, 11745–11751.
47. D. N. Congreve, J. Lee, N. J. Thompson, E. Honz, S. R. Yost, P. D. Reusswig, M. E. Bahlke, S. Reineke, T. Van Voorhis, M. A. Baldo, *Science*, 2013, 340, 334–337.
48. F. Bella, G. Griffini, J.-P. Correa-Baena, G. Saracco, M. Grätzel, A. Hagfeldt, S. Turri, C. Gerbaldi, *Science*, 2016, 354, 203–206.
49. H. Vogel, *Chem. News*, 1873, 318–319.
50. B. O'Regan, M. Grätzel, *Nature*, 1991, 353, 737–740.
51. A. Yella, H.-W. Lee, H. N. Tsao, Ch. Yi, A. K. Chandrian, M. K. Nazeeruddin, E. W.-G. Diau, Ch.-Y. Yeh, S. M. Zakeeruddin, M. Grätzel, *Science*, 2011, 334, 629–634.
52. A. Kojima, K. Teshima, Y. Shirai, T. Miyasaka, *J. Amer. Chem. Soc.*, 2009, 131, 6050–6051.
53. M. M. Lee, J. Teuscher, T. Miyasaka, T. Murakami, H. J. Snaith, *Science*, 2012, 338, 643–647.
54. S. Yang, W. Fu, Z. Zhang, H. Chen, Ch.-Z. Li, *J. Mater. Chem. A*, 2017, 5, 11462–11482.

55. R. Prsanna, A. Gold-Parker, T. Leijtens, B. Conings, A. Babayigit, H.-G. Boyen, M. F. Toney, M. D. McGehee, *J. Amer. Chem. Soc.*, 2017, 139, 11117–11124.
56. M. V. Kovalenko, L. Protesecu, M. I. Bodnarchuk, *Science*, 2017, 358, 745–750.
57. M. Saba, *Nature*, 2018, 553, 163–164.
58. M. A. Becker *et al.*, *Nature*, 2018, 553, 189–193.
59. S. Rühle, M. Shalom, A. Zaban, *ChemPhysChem*, 2010, 11, 2290–2304.
60. N. A. Campbell, J. B. W. Reece, L. A. Urry *et al.*, Biology, 8th ed., Pearson, Benjamin Cummings, San Francisco 2008.
61. W. S. Fyfe, H. Puchelt, M. Taube, *The Natural Environment and the Biogeochemical Cycles*, ser. *The Handbook of Environmental Chemistry*. Springer, Berlin Heidelberg, 2013.
62. N. S. Lewis, D. G. Nocera, *Proc. Natl. Acad. Sci. USA.* 2006, 103, 15729–15735.
63. J. Barber, P. D. Tran, *J. R. I Soc. Interface*, 2013, 10, 1–16.
64. R. E. Blankenship, D. M. Tiede, J. Barber, G. W. Brudvig, G. Fleming, M. Ghirardi, M. R. Gunner, W. Junge, D. M. Kramer, A. Melis, T. A. Moore, C. C. Moser, D. G. Nocera, A. J. Nozik, D. R. Ort, W. W. Parson, R. C. Prince, R. T. Sayre, *Science*, 2011, 332, 805–809.
65. R. J. Cogdell, T. H. P. Brotosudarmo, A. T. Gardiner, P. M. Sanchez, L. Cronin, *Biofuels*, 2010, 1, 861–876.
66. P. Fromme, *Photosynthetic Protein Complexes: A Structural Approach*, Wiley-Blackwell, Weinheim, 2008, p. 360.
67. R. E. Blankenship, *Molecular Mechanisms of Photosynthesis*, Blackwell Science Ltd., Oxford, 2014.
68. Z. Liu, H. Yan, K. Wang, T. Kuang, J. Zhang, L. Gui, X. An, W. Chang, *Nature*, 2004, 428, 287–292.
69. V. I. Novoderezhkin, M. A. Palacios, H. van Amerongen, R. van Grondelle, *J. Phys. Chem. B*, 2005, 109, 10493–10504.
70. T. P. J. Krüger and R. van Grondelle, *Physica B: Condensed Matter*, 2016, 480, 7–13.
71. G. D. Scholes, G. R. Fleming, A. Olaya-Castro, R. van Grondelle, *Nat. Chem.*, 2011, 3, 763–774.
72. R. van Grondelle, V. I. Novoderezhkin, *Nature* 2010, 463, 614–615.

73. B. Gobets B, R. van Grondelle, *Biochim. Biophys. Acta — Bioenergetics*, 2001, 1507, 80–99.
74. H. Frauenfelder, S. G. Sligar, P. G. Wolynes, *Science*, 1991, 254, 1598–603.
75. G. Weber, *Adv. Protein Chem.*, 1975, 29, 1–83.
76. F. Fassioli, R. Dinshaw, P. C. Arpin, G. D. Scholes, *J. R. Soc. Interface*, 2014, 11, 0130901.
77. M. Wahadoszamen, I. Margalit, A. M. Ara, R. van Grondelle, D. Noy, *Nat. Commun.*, 2014, 5, 5287.
78. A. A. Pascal, Z. F. Liu, K. Broess, B. van Oort, H. van Amerongen, C. Wang, P. Horton, B. Robert, W. R. Chang, A. Ruban, *Nature*, 2005, 436, 134–137.
79. A. V. Ruban, R. Berera, C. Ilioaia, I. H. M. van Stokkum, J. T. M. Kennis, A. A. Pascal, H. van Amerongen, B. Robert, P. Horton, R. van Grondelle, *Nature*, 2007, 450, 575–578.
80. C. Ramanan, J. M. Gruber, P. Maly, M. Negretti, V. I. Novoderezhkin, T. P. J. Krüger, T. Mancal, R. Croce, R. van Grondelle, *Biophys. J.*, 2015, 108, 1047–1056.
81. T. P. J. Krüger, P. Maly, M. T. A. Alexandre, T. Mancal, C. Büchel, R. van Grondelle, *Proc. Natl. Acad. Sci. USA* 2017, 114: E11063–E11071.
82. T. P. J. Krüger, R. van Grondelle, *J. Phys. B: At. Mol. Op. Phys.*, 2017, 50, 132001.
83. T. P. J. Krüger, C. Ilioaia, M. P. Johnson, A. V. Ruban, E. Papagiannakis, P. Horton, R. van Grondelle, *Biophys. J.*, 2012, 102, 2669–2676.
84. L. Hammerstöm, *Chem.*, 2016, 1, 515–519.
85. E. Roduner, *Catal. Today*, dx.doi.org/10.1016/j.cattod.2017.05.091.
86. E. Skúlarson, G. S. Karlberg, J. Rossmeisel, T. Bligaard, J. Greeley, H. Jónson, J. K. Nørskov, *Phys. Chem. Chem. Phys.*, 2007, 9, 3241–3250.
87. Y. Zheng, Y. Jiao, Y. Zhu, L. Hua, Y. Han, Y. Chen, A. Du, M. Jaroniek, S. Z. Qiao, *Nat. Comm.*, 2014, 5, 3783.
88. I. Roger, M. A. Shipman, M. D. Symes, *Nat. Rev. Chem.*, 2017, 1, 1–13.
89. Y. -C. Nah, I. Paramasivam, P. Schmuki, *ChemPhysChem*, 2010, 11, 2698–2713.
90. M. Grätzel, *Nature*, 2001, 414, 338–344.

91. K. Maeda, K. Domen, *J. Phys. Chem. Lett.*, 2010, 1, 2655–2661.
92. K. Maeda, A. Xiong, T. Yoshinaga, T. Ikeda, N. Sakamoto, T. Hisatomi, M. Takashima, D. Lu, M. Kanehara, T. Setoyama, T. Teranishi, K. Domen, *Angew. Chem. Int.*, 2010, 49, 4096–4099.
93. J. Yang, D. Wang, C. Li, *Acc. Chem. Res.*, 2013, 46, 1900–1909.
94. J. H. Montoya, L. C. Seitz, P. Chakthranont, A. Vojvodic, T. F. Jaramillo, J. K. Nørskov, *Nat. Mat.*, 2017, 16, 70–81.
95. N. Cox, J. Messinger, *Biochim. Biophys. Acta*, 2013, 1827, 1020–1030.
96. R. Brimblecombe, D. R. J. Kolling, A. M. Bond, G. Ch. Dismukes, G. F. Swiegers, L. Spiccia, *Inorg. Chem.*, 2009, 48, 7269–7279.
97. H. Park, W. Choi, M. R. Hoffmann, *J. Mat. Chem.*, 2008, 18, 2379–2385.
98. J. Tisler, G. Balasubramanian, B. Naydenov, R. Kolesov, B. Grotz, R. Reuter, J.-P. Boudou, P. A. Curmi, M. Sennour, A. Thorel, M. Börsch, K. Aulenbacher, R. Erdmann, P. R. Hemmer, F. Jelezko, J. Wrachtrup, *ACS Nano*, 2009, 7, 1959–1965.
99. R. Hanson, V. V. Dobrovitski, A. E. Feiguin, O. Gywat, D. D. Awschalom, *Science*, 2008, 320, 352–355.
100. G. Balasubramanian, P. Neumann, D. Twitchen, M. Markham, R. Kolesov, N. Mirzuochi, J. Isoya, J. Achard, J. Beck, J. Tissler, V. Jacques, P. R. Hemmer, F. Jelezko, J. Wrachtrup, *Nature Mat.*, 2009, 8, 383–387.

Index

9 781800 616349